OPTOKINETICS

OPTOKINETICS

A NEW SYSTEM OF OPTICS

HARRY H. MARK

LIBRARY OF CONGRESS CONTROL NUMBER: 2011910991
ISBN: HARDCOVER   978-1-4628-9737-7
                       SOFTCOVER   978-1-4628-9736-0
                       EBOOK   978-1-4628-9738-4

This book was printed in the United States of America.

To order additional copies of this book, contact:
Xlibris Corporation
1-888-795-4274
www.Xlibris.com
Orders@Xlibris.com
95411

CONTENTS

DEFINITIONS

To my daughters Tami and Hayley, partners in search of truth.

ίητρος φιλόσοφος ίσόδεος
—Hippocrates

Preface

This volume is a slightly enlarged and amended edition of my treatise by the same name published in 1982. My system does not deal with the long-standing dilemma concerning the actual nature of light, whether it's a material corpuscle, a wave in a material ether, or a mental construct in mathematical terms, which combines the two. Optics is a branch of physics, and physics nowadays is wedded to mathematics, whereas I am impressed by quantitatively perceptible reality advocated by Ernst Mach and Max Planck: "Physics is an exact science and hence depends upon measurement, while all measurement itself requires sense-perception."[1] I return in this way to where Newton has left off in his *Opticks*, or to paraphrase Justus Liebig: The progress of science is like a winding staircase which by the end of each turn arrives close to the starting point, but hopefully a bit higher.

Some of my main points of divergence from the orthodox system are in the treatment of reflection (Newton's axiom II), reciprocity (Newton's axiom III), and refraction. Optical phenomena are here interpreted by the one measurable physical property common to all lights—their motions. I believe that the crux of the problem in the old system may have been its difficulty when dealing with color (monochromatism) to distinguish between complicated physiological perceptive information from simple quantitative physical stimuli, even though at least two prominent contributors to the system, Thomas Young and Hermann Helmholtz were medical doctors exceedingly well versed in the mechanism of the eye. As shown by Johannes Müller almost two centuries ago, different stimuli can produce the same sensations.[2] In addition, psychosocial factors may have delayed advances beyond the orthodox system such as the inordinate idolatry of Isaac Newton after the success of his *Principia* and of Albert Einstein after being granted the Nobel Prize for explaining the photoelectric effect (objected to by Nobel laureate in medicine Allvar Gullstrand).

For a number of well-known reasons, I harbor no illusions that my system will be seriously considered anytime soon. First, the simple experiments that form the foundation of the theory must be, as a matter of course, independently duplicated, which may not be difficult. Secondly, the predictions of the theory concerning the speeds of lights must be realized, which will probably be hard to do with present technology. And thirdly, as was once said, before a new paradigm or theory is considered the generation that was brought up on the time-honored one must pass away. For my part, I felt it my civic duty before I depart this world to share with the public my personal experiences and ideas just in case they prove of some benefit to the advancement of knowledge. Charles Darwin has expressed my sentiments in this regard much better than I can ever hope to do:

> Many of the views which have been advanced are highly speculative, and some no doubt will prove erroneous; but I have in every case given the reasons which have led me to one view rather than to another. False facts are highly injurious to the progress of science, for they endure long; but false views, if supported by some evidence, do little harm, for everyone takes a salutary pleasure in proving their falseness; and when this is done, one path towards error is closed and the road to truth is often at the same time opened.[3]

1. Planck M. The Universe in the Light of Modern Physics. New York, WW Norton; 1931: 7.
2. Kuhn, TS. The Structure of Scientific Revolutions. Chicago. Univ. of Chicago Press; 1970: 192-3.
3. Darwin C. The Descent of Man. New York, AL Burt, 1874: 693.

OPTOKINETICS

Refraction

INTERFACE EVENTS

Light crossing obliquely between air and water changes direction; it is bent, fractured, refracted. The phenomenon was recorded early in history in the broken appearance of an oar partially immersed in water. Since early in the Renaissance, refraction received close attention because it was found to form the basis of our acute vision and the production of lenses—for eyeglasses, field glasses, telescopes, microscopes, and most other instruments meant to aid the eye to see better, farther, and larger. Refraction lies at the foundation of almost all our knowledge inasmuch as almost all our knowledge is obtained visually.

The fact that refraction occurs between transparent bodies—like air, water, or glass—offers a unique opportunity for easy inspection and, perhaps, reflection. In the following pages, we shall inquire into how and why this refraction happens.

The term *optical medium* refers to a homogeneous substance through which light passes to an appreciable distance and which is therefore transparent. Vacuum, though not a true substance, is considered a medium, but iron is not, even though in thin sheets it transmits light. *Interface* denotes the boundary between media and is considered smooth and flat. The interface may appear to the unaided eye as a well-defined plane or a line in cross section, but at higher magnification (on a molecular level), it becomes ever harder to delineate. Correspondingly, the exact location of interface events is somewhat ambiguous. Does the change in light's course by refraction occur before reaching the interface or after crossing it? Is light reflected prior to meeting the second medium, or after entering it, as Newton thought? These questions are relevant when inquiring deeper into the phenomena where the exact nature of the abstract geometrical line becomes interesting. They are, however, beyond the scope of this discussion.

Available evidence indicates that events at the interface occur mostly on one plane and may, up to a certain degree of accuracy, be represented in two dimensions; light moving in the plane of this page does not refract or reflect toward the reader or into the page. Other times, the third dimension must be considered.

In addition to the dimensions of space in their geometrical representation, interface phenomena possess the dimension of time, for no real event occurs without the passage of time. From Ptolemy in antiquity to Descartes in the Renaissance, the belief held that light propagated instantly. Accordingly, since it took no time for incident light to be, say, reflected or refracted at the interface, it was not a true event; and static Euclidean geometry alone, operating deductively from axioms, sufficed to elucidate the phenomena. The certainty of the method made optics the queen of all sciences.

The angles of incidence and reflection had to be equal in order to subtend the geometric minimum distance, tacitly assuming also that nature or the Creator did nothing in vain—a teleological principle of economy popular from antiquity to this day. The premise in this line of syllogistic reasoning was removed when, later, light was discovered to propagate in time and the equality of the angles of incidence and reflection could no longer be validly deduced from it. The conclusion may or may not be true, but it certainly did not anymore follow from the premise, which had been proven false.

When ordinary light strikes an interface obliquely, at least four events take place:

1. Light arriving from one medium is reflected back into this medium.
2. Light crosses the interface into the second medium and changes its course by refraction.
3. This transmitted light is somewhat diffused, dispersed, in the second medium.
4. Colors of the spectrum are often perceived.

The events may be treated separately, one by one for instance, by employing monochromatic light to reduce the sometimes annoying colored dispersion, though this convenience is paid for by diminished insight into the nature of color; and events at the interface remain, to that extent, less completely understood.

Integrated treatment of interface events not only has the advantage of comprehensiveness but is dictated by the events themselves, which occur all at the same place at the same time. Some correlation or qualitative interdependence may therefore be validly suspected according to John Stuart Mill's rule of concomitant variations: Whatever phenomenon varies in any manner whenever another phenomenon varies in some particular manner, is either a cause or an effect of that phenomenon, or is connected with it through some fact of causation.[1, 2] A phenomenon that appears dependent on another is traditionally designated as being the function of the other. The degree of refraction, for instance, is a function of light's speed in the transparent media.

Proceeding with this method, we first simply note the circumstances attended to each interface event that enters into the function of refraction (analyze the phenomenon) and, secondly, integrate these pieces of information to see if we are thus led to correlations that enhance its understanding. Equipped with this information, derived as best possible from easily verifiable perceptual data, we then view other optical phenomena in which refraction plays a role, such as dispersion.

Nomenclature

Names carry connotations that embody the prejudices of the original nominators and which, therefore, subliminally influence their meaning perceived by the reader. In order then to approach a subject unbiased, it is important to glance for a moment at the meanings hidden behind the terms.

The angle of incidence traditionally denotes the angle that a beam of light forms with the *normal* (perpendicular) to the interface it strikes. The choice of the normal to define the angle, rather than the interface itself, assumes that the geometric normal is a properly firm term by which to define a real phenomenon in nature. It reflects the confidence the nominators had in Euclidean geometry, which seemed to them best suited to help arrive at reliable knowledge of nature. While it was satisfying to see nature relating in this manner to the human intellect, its elucidation became complicated, uncertain, and removed from reality itself, commensurate with the distance between the abstract idea and actual experience. For

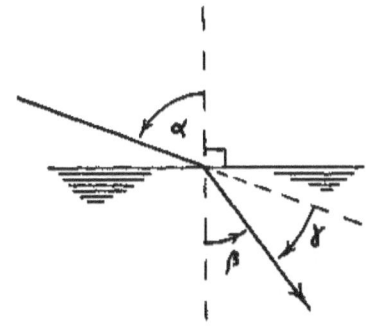

these reasons, a return to Kepler's angle of inclination is preferable: the smaller of the two angles formed by an incident beam with the interface. It equals 90 - α, where α denotes the traditional angle of incidence.

The angle of refraction designates the angle in the second medium that a refracted beam of light forms with the normal to the interface—β (figure). The choice is again unfortunate, for as the phenomenon of refraction consists of a beam changing its course, it is preferable to name the actual physical event—the *deviation* from the course γ—rather than its relation to an abstract geometrical concept. We may do so by emphasizing the angle of deviation—the angle that a beam, after passing the interface, forms with its original direction before reaching this interface. Kepler indeed named this angle the angle of refraction.

The magnitude of the angle of deviation—the refraction—varies with the nature of the medium and with the nature of light. The ease with which a medium refracts light was once termed *refrangibility* and, today, its *refractive index*, *refractive coefficient*, or *refractivity*. Glass is more refractive than water; its refractivity is higher. The same words denote the ease with which a light is refracted. Violet appearing light is more refractive than red; its refractivity is higher.

Light passing obliquely from the rarer medium into the denser one is traditionally termed as deviating toward the normal (γ); when it passes in the other direction, its deviation is away from the normal (α). One may thus be led to believe that whatever causes the deviation in the two transitions results in opposite effects—toward versus away—where the one event was perhaps the reciprocal of the other. Viewed, however, with regard to real physical objects rather than abstract geometry, light that arrives at the interface from the side of either medium is deviated in only one direction—toward the denser medium (see "Dispersion").

Parameters of Refraction

1. THE ANGLE OF INCIDENCE

Early on, refraction was examined systematically in the catoptrics attributed to Euclid (300 BC).[3] A hidden object (D) in an empty vessel BGDE becomes visible to an observer (A) standing aside it when the vessel is filled with water. Ptolemy (AD 150) recognized that light from the submerged object must bend at the interface in order to reach the observer. He proceeded to inquire what this angle was by numerous measurements, later tabulated, and concluded that for any two media, the angle of refraction was proportional to the angle of incidence: α/β = constant.[4] The idea of direct proportionality persisted into the Renaissance so, for instance, "angles of incidence are proportional to angles of deviation"[5] (Franciscus Maurolicus, 1494-1575). Empirical tables of refraction for air, water, wine, oil, or glass were prepared by others after Ptolemy, notably Alhazen (1000), his Latin translator Vitello (1270), and Kircher (1671).

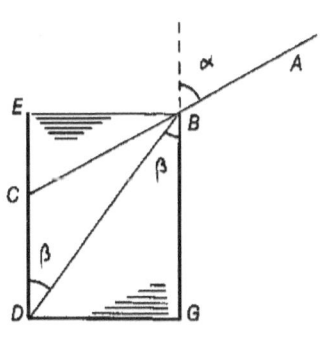

When Johannes Kepler (1571-1630) looked at these and his own tables, he recognized that no direct proportionality existed and, indeed, could not exist when he considered the entire range of incidences from 0° to 90° and the corresponding range of refracted rays on either side of the interface.[6] When the angle of incidence in the rarer medium was gradually changed from the perpendicular (incidence 0°) to nearly grazing the interface, the refracted ray never did reach this interface. For every angle of incidence, the angle of refraction was smaller.

Kepler finally concluded that the angle of deviation was composed of two parts, one proportional to the angle of incidence and the other proportional to the secant of the angle of refraction.[7] Calculated according to this formula, his results agreed more closely with actual measurements than those of his predecessors.

A few years later, Willebrord Snell (1591-1626) continued the pursuit but did not publish his work so that his mode of reasoning, his theory and explanation, perished with him; the results were fortunately known to others, such as Huygens and Descartes, who duly recorded them. Observing the course of incident rays at various angles to interface BE (figure), Snell noted that for any pair of media, particularly air and water, the length of the refracted ray BD related to the length of the unrefracted direction BC as the cosecants of the angle of refraction to that of incidence: $BD/BC = \cosec\beta/\cosec\alpha$ = constant. Since cosecant = 1 /sinus, one wrote $\sin\alpha/\sin\beta$= constant, and this is *Snell's law of refraction* as first published by Descartes.[8] When the rarer medium is air (or vacuum), the constant of the equation is termed the refractive index (n) of the denser medium and is normally greater than 1.

Snell's equation represented an enormous achievement, for it was thereafter possible to reliably predict for every angle of incidence the direction of the refracted ray by measuring these angles only once for each medium. All the tables of refractive angles laboriously assembled in over a millennium became, in one sweep, obsolete. It greatly eased the construction of the newly invented telescopes and microscopes and all other optical instruments to this day. Having been accepted as the final solution to the phenomenon of refraction, the formula may have, at the same time, discouraged its further exploration.

What the trigonometrical ratio $\sin\alpha/\sin\beta > 1$ says is that at a small angle of incidence α, near the perpendicular, the angle of refraction β is also small, as is the angle of deviation (α-β). But as α grows, β grows slower (as the sines of small angles grow slower than those of large ones) and α-β increases, that is, the deviation—the refractivity—is a function of the angle of incidence.

Historians sometimes wondered why this seemingly simple relationship eluded the giant Kepler, upon whose shoulders young Snell stood. Apparently, Kepler's sights were higher; he seems to have pursued not merely a quantitative relationship but also the *physical cause*: what made light deviate at the interface in the manner it did?

Alhazen[9] reasoned teleologically that light sought to recover its loss caused by the obliquity of incidence. The more it was weakened by this distance from the normal, the more it would recover by nearing the normal, for vertical light, so he said, was the strongest; hence the deviation in the denser medium was toward the vertical. Kepler similarly wrote that the resistance of the medium increased with the angle of incidence—as the angle increased so did the cross-section surface area upon which a light beam of given width fell, i.e., there was more of the denser medium in its way that caused more resistance and, hence, more refraction. The enlarged cross section was, however, not

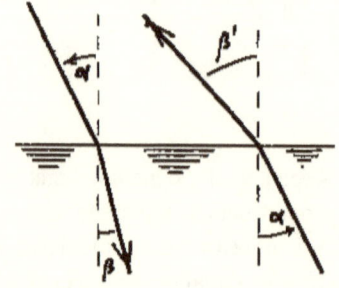

directly proportional to the angle of incidence, and Kepler referred the reader to the table of secants for a better representation of this proportionality.

When the direction of incident light in the *rarer* medium is shifted from the perpendicular to a certain angle α, the refracted light in the denser medium shifts less—from the perpendicular to angle β. On the other hand, when incident light in the *denser* medium is displaced by the same magnitude α, the shift of the refracted light is much larger, $\beta'>\beta$. For every increment of angular displacement of incidence, there is a larger angular displacement of deviation when light transits in direction denser-to-rarer medium than in the opposite direction (figure). Or in other words, the refractivity (the facility of refraction) is higher when incident light arrives at the interface from the side of the denser medium than when it arrives from the side of the rarer one.

The crucial question is, why do the angles vary as their sines? Or more precisely, for identical incremental changes in the angle of incidence between 0° and 90°, why does the deviation by refraction vary, and why does it vary from one side of the interface to the other? Snell's empirical formula did not answer Kepler's question. It was bound to have its limitations.

One of these limitations, discovered and dealt with at length by Kepler, concerns refraction out of the denser medium. When the angle of incidence here exceeds a certain value—the critical angle—no refraction at all takes place, but rather light is only reflected—total reflection.

Why does light suddenly cease to penetrate the interface? Kepler admitted his loss to understand it and added, "Give me the answer, wherever you live, and you shall be as Apollonius to me." Mach wrote that "one must not come to the conclusion given in many textbooks: If $\sin\beta = \sin\alpha/n$, and $n>1$, it follows that, when $\sin\alpha > n$, $\sin\beta > 1$, an impossible value, *hence* [sic] there is no refraction. On the contrary, it only shows that the formula does not represent this case any more."[10]

Evidently, in order to study the phenomenon, other than mathematics was needed, for nature does not usually abide by the limitations of mathematical logic. What is it physically in the angle of incidence that causes light to change its course in Snell's manner? Based on empirical observations, we may, at this juncture, merely accept three facts:

1. The angle of deviation is a function of the angle of incidence—refractivity and the angle of incidence are in some way proportional.
2. This proportionality is not direct and linear, for refractivity varies more than the angle of incidence.
3. Refractivity is greater on denser-to-rarer transit than in the opposite direction.

2. VELOCITIES

An empirical formula, or an isolated phenomenon, does not usually satisfy the human mind, ever seeking reasons, causes, and explanations to fit it within a rational order. Hypotheses to explain Snell's law were not long in coming. Curiously, the first to introduce to the discussion of refraction a physical parameter—that of velocity—was René Descartes (1596-1650), who curiously maintained at the same time that light propagated instantaneously without velocity. He believed space to be filled with small spinning bodies and denied the possibility of a vacuum, the reality of which was demonstrated the year of his death by Guericke with his air pump. To Descartes, light was "nothing other than a certain motion or an action conceived in a very subtle matter,"[11] analogous to a strike at one end of a rod that was instantly received at the other end. Accordingly, light moved faster in stiffer, denser media than in rarer, softer ones. "Thus it happens that so much as the small parts of a transparent body are harder and firmer, so much the more do they allow the light to pass more easily."

Descartes, who had great faith in geometry, drew a circle with center C (figure), bisected by interface DCE, and said that the speed vector of incident light (AC) is divided at the interface in direction of the normal CG and in direction of the surface CE (CG >CE). The vector in direction CG is greater than CE, and hence light deviates toward the normal to B. The relation of AH to BG, said Descartes, is constant for any two media, which means sinα/sinβ=const.

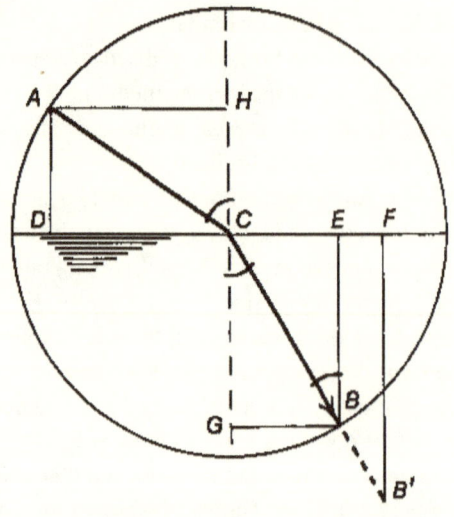

The lawyer-mathematician Pierre Fermat (1601-1665) provided a firmer logical foundation to Snell's formula by applying Hero's teleological principle of least time (see "Reflection") to his newly invented method of maxima and minima—a precursor of differential calculus. In 1662, Fermat said, "Naturam operari per modos faciliores et expeditiores. Naturam per lineas brevissimas semper operari" (nature always operates along the shortest lines).[12] Contrary to Descartes, Fermat intuitively and without evidence presupposed that the velocity of light in the denser medium was lower than in the rarer one. "The resistance of the rarer medium ought to be less than that of the denser medium, by an axiom which it is natural to adopt."[13]

The question may be stated by analogy to a man in position A, say, in New Haven, Connecticut, who wishes to travel at maximum speed limit and is straight lines to position B, say, New York City, New York. The speed limit in Connecticut v_1 is higher than in New York v_2. Where on the state line DCE shall he cross? Fermat answered (as did Descartes, with whom he corresponded at length on the subject) by drawing a circle with center c (figure) bisected by interface DE and then deducing where point B is located on the circle so that travel time AB was minimum. We designate $DC = x$, $DE = e$, $AD = h_1$, and $EB = h_2$. Time is distance per velocity ($T = D/v$), hence, $AC/v_1 + CB/v_2 = $ Minimum.[8]

$$\frac{\sqrt{h_1^2 + x^2}}{v_1} + \frac{\sqrt{h_2^2 + (e-x)^2}}{v_2} = Min.; \quad \frac{dM}{dx} = 0$$

$$\frac{x}{v_1\sqrt{h_1^2 + x^2}} - \frac{e-x}{v_2\sqrt{h_1^2 + x^2}} = 0;$$

$$\sin\alpha/vi - \sin\beta/v_2 = 0; \quad \sin\alpha/\sin\beta = n = V_1/V_2.$$

The marvel of mathematics is such that although Fermat arrived at the same sine law as Descartes, their suppositions as to the velocity of light were diametrically opposed. Where Descartes held it higher in the denser medium, Fermat held it lower. Both theories, both modes of logical proof were valid, but the basis of the law remained empirical. Real measurements of the angles seemed to support Snell's equation.[14]

Descartes's and Fermat's proofs were arbitrarily limited to the case in which the starting and end points were on one circle bisected by the interface, circumstances that rarely occur in nature. Suppose that light, instead of wishing to travel to B, elects to go to B'. Its path through the slower medium is extended and the time to cover it prolonged. To compensate for it in the spirit of least time, this path must be shortened (made more perpendicular to the interface), i.e., the point of transit C moves toward F. The farther point B' is from the interface, the closer to the normal the line CB', and the greater the deviation. It follows that the angle of refraction is a function of the distance of point B' from the interface, which is contrary to fact.

A simple explanation of refraction was then furnished in 1674 by the famous physicist Claudius Franciscus Milliet Deschales (1621-1678) in his monumental (3,000 pages) book (where, among others, he gave the first correct interpretation to floaters in the eye). He considered light not merely a linear ray, as was usually portrayed, but a beam with a front (figure). As each successive front entered the medium on one of its sides, this side slowed down and, thus, the whole front deviated toward the normal to the interface.[15]

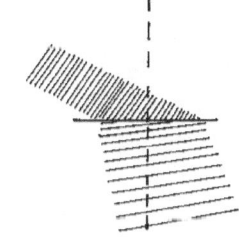

Some years later, Christiaan Huygens (1629-1695) expounded on Deschales's idea by equating the beam's front to the front of a water wave. To him, as to Aristotle, light was not something in motion, but a force that acted on something, the "ethereal aether" of antiquity. The action caused the particles of this matter to move in an undulatory fashion similar to particles of water that are moved by the action of a falling stone. "It is inconceivable to doubt that light consists in the motion of some sort of matter. Light consists in a movement of the matter which exists between us and the luminous body."[16]

Huygens argued, contrary to Descartes, that the undulatory motion of light was slower in denser media than in air. "If the particles of transparent bodies have a recoil a little less prompt than that of the ethereal particles, which nothing hinders us from supposing, it will again follow that the progression of the waves of light will be slower in the interior of such bodies than it is outside in the ethereal matter."

At the time it takes point B (figure) of incident light AB to reach the interface AC, point A arrives somewhere on the semicircle EDF with radius AD representing the scalar magnitude of the velocity in the denser medium. The tangent to the semicircle through point C locates point D, thus providing the direction of the vector whereto light is refracted. $BC/V_1 = AD/V_2$. $BC = AC \sin\alpha$, $AD = AC \sin\alpha$ and, hence, $\sin\alpha/V_1 = \sin\beta/V_2$, or $\sin\alpha/\sin\beta = V_1/V_2$. This mode of reasoning was often named Huygens's Construction. The semicircles that originate at the interface and whose radii correlate to the velocity of light in the medium was later named Huygens's principle.[17]

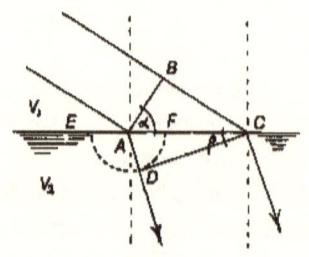

Huygens reasoned by analogy when he maintained that light striking the interface at A creates a wave with front EDF (other such semicircles originated at each point along the line of impact). If one has difficulties conceiving the semicircle EDF as a sealer magnitude, one may materialize it by equating this page with the surface of a pond upon which a stone dropped at point A; the semicircle then represents the position that the ensuing wave attained a given interval of time following its creation. Not everyone had such conceptual difficulties and, hence, not everyone needed the material analogy of the wave. As Sabra pointed out, there was nothing in Huygens's illustration to compel the idea of a wave.[18] He did not set out to determine the law of refraction, which he already had from Snell, but rather to seek approval of his new concept of light waves by the fact that he could explain reflection and refraction by their means, just as Fermat sought approval of his mathematical method by showing that it could explain refraction.

Finally, Isaac Newton (1642-1727) was understood to have held that light was composed of "globules" that obeyed his laws of gravitational attraction. Accordingly, these particles were attracted by the heavier, denser medium and were refracted toward it, or toward the normal to the interface, while, at the same time, their velocity increased by this force.

All attempts at deducing light's velocities from Snell's equation led to contradictory results. When, in 1850, Léon Foucault (1819-1868) actually measured that velocity in water and found it lower than in air,[19] it enhanced those theories that counted on it—it almost divined Fermat's principle of least time or Huygens's notion of wave motion—while detracting from Newton's idea that gravity influenced light.

The fact that light travels slower in denser media did not by itself explain refraction because the ratio of velocities between two given media is always constant—also at perpendicular incidence when there is no refraction at all. Neither did it explain why the angle of *deviation* increases with the angle of incidence.

Following customary principles of logic, we note first that for a *constant angle* of incidence the refractivity varies when the velocity in the second medium varies. On the other hand, for a *constant velocity* in the second medium, the refractivity varies when the angle of incidence varies. The same effect is obtained by two different causes: the effect caused by a growing angle of incidence is the same as that caused by slower velocity in the second medium (a higher velocities ratio). The angle is, however, a geometrical term. At this point, we still do not know what it means in physical reality. What happens to light at the interface when the angle changes? We may only suspect that an increase in the angle has something in common with reduced velocity in the second medium because they both result in the same refractive effect.

For one and the same angle of incidence, light that arrives at the interface from the side of the denser medium, where its velocity is lower, is more deviated than when it arrives with the higher velocity from the rarer medium.

In summary, we may then say that the higher the ratio of velocities at the interface, the greater the refraction; and for a given ratio, the refraction is greater when light arrives from the side where it travels slower. Furthermore, refraction may be altered either by a change of velocity in one of the media or by a change in the angle of incidence.

3. DISPERSION

Dispersion refers here to a phenomenon attended to refraction whereby incident light after refraction becomes dilated and fans out. Ordinary light transmitted through a wedge-shaped piece of glass (a prism) is refracted and dispersed to form a colored spectrum. The order of colors in it proceeds from red at one extremity to orange, yellow, green, blue, and finally violet at the other extremity. The spectrum is in this sense, and without prejudice, polarized; it has a red pole and a violet pole.

The spectrum had been known since antiquity and received particular attention by Newton, who emphasized that the red pole was consistently the least deviated by refraction and the violet pole the most deviated. Furthermore, when Newton separately transmitted one colored segment through a prism, or any sequence of prisms, he found that it continued to be more or less deviated than the other segments, according to its geometrical position in the spectrum. Consequently, Newton generalized that every color had its own specific and unalienable refractivity.

The concept represented a very substantial advance because colors—hitherto regarded as abstract aesthetic sensations, mere qualities—were thereby given concrete properties that could be quantitatively measured. By observing the geometrical angle of refraction even a color-blind person could distinguish them. Elaborating on this act of quantification took another century and another theory.

The variable refractivity of colors introduced yet an additional constraint to the validity of Snell's formula. In 1672, Newton called the formula a hypothesis;[20] though in 1704, he cautiously wrote it as an axiom, "The Sine of Incidence is either accurately or very nearly in a given Ratio to the Sine of Refraction," but added the restriction of color, "The Sine of Incidence of the red Light is to the Sine of its Refraction as 4 to 3. In Light of other Colours the Sines have other Proportions.[21] Modern texts say, "Well-known experiments on prismatic colours, first carried out by Newton, show that the index of refraction depends on the colour."[22]

Applied to the refraction of colors, Snell's law—as finally formulated by Huygens and Fermat—means that the velocities of the various colors in denser media vary; $\sin\alpha/\sin\beta = V_1/V_2$ means that for a given a and vi, the greater the deviation, the lower the velocity in the denser medium; and since the violet end of a spectrum is the most deviated, the velocity of violet is the lowest.[23]

Actual data on these velocities are as yet hard to find in the literature. Moreover, assuming the velocities of all colors in vacuum or air to be the same, it follows that for a common angle of incidence the violet suffers a greater loss of velocity in the denser medium than the red; in order to equalize the refraction of red to that of violet, the velocity of red in the second medium must be reduced.[24] Similar as the angle of deviation by refraction grows with increasing angle of incidence, from 0° at perpendicular incidence to a maximum at grazing, so does the

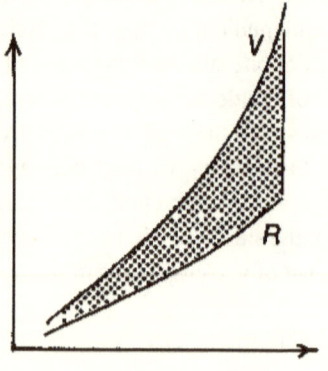

angle of dispersion increase with the angle of incidence. The distance between the red end of the spectrum and its violet end increases with the angle of incidence (figure). Furthermore, and again similar to the angle of deviation, the increase in dispersion is not linear, i.e., it grows ever faster, as seen in the figure, where on the abscissa is plotted the angle of incidence and on the ordinate the corresponding angle of deviation for the various colors. The curve represents the *normal dispersion*.

In the next figure, similar to Huygens's construction, incident beam AB is refracted and dispersed in directions AEF and CD. By the time it takes light v from E to arrive at F, light R goes from C to D. Since CR is at an angle to AV, the distance CD is longer than EF, i.e., light moves faster. As the angle of incidence is now increased, the angle between AV and CR increases (dispersion) and with it the ratio DC/EF. This means that as the angle of incidence increases, the velocity of violet, for instance, decreases more than the velocity of red. In order to equalize the velocities, red must be more retarded when the angle of incidence is high than when it is low.

Let a certain glass have a refractive index of, say, $n = 1.6170$, for the red end of a spectrum, and $n = 1.6350$ for the violet end, $V_1/V_2 = n$. If the speed in air V_1 is taken to be, say, 300,000 km/sec for both colors, it follows that in the glass V_2 red = 185,500 km/sec and V_2 violet = 183,500 km/sec, a difference of 2,000 km/sec. If these two segments of one spectrum are transmitted through 2,000 km of glass, the red will emerge one second before the violet. Allowing Newton's idea of white light as a loose aggregate of colors, white light thus viewed through glass must, in time, change colors from red to violet.

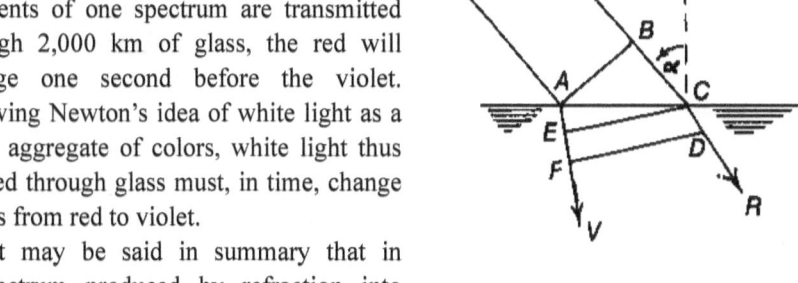

It may be said in summary that in a spectrum produced by refraction into a denser medium, the various color-evoking lights travel at various velocities, and these velocities diminish in nonlinear fashion from the red end to the violet end of the spectrum. In addition, dispersion varies nonlinearly with the angle of incidence so that the difference in the deviations of the spectrum's extremes (red-violet) grows at an accelerated rate with simple growth of the angle of incidence.

4. REFLECTIVITY

Light striking an interface is divided into two parts, one is returned into the first medium while the other traverses into the second medium. From a smooth interface, as we assume here, the reflection is denoted specular, distinct from diffuse scattering from rough surfaces. The ease with which reflection occurs was once termed reflexibility, and in modern usage reflectance, reflection coefficient, or reflectivity. It determines the fractional portion of incident light that is reflected.[24] The ease of penetration through the interface is termed transmittance, or transmissivity. The total amount of light after an encounter with an interface is made up of the reflected and transmitted parts, the more reflected the less transmitted—the higher the reflectivity the lower the transmissivity and vice versa.

The fact that a given quantity of light, moving uniformly in a rarer medium, does slow down after transit through an interface leads to the conclusion that it either becomes more concentrated in the denser medium (higher intensity, which does not happen) or else some of it remains in the rarer one. The slower the motion in the denser medium (or the greater the ratio of velocities, i.e., the refractive index), the more light must be reflected back into the rarer medium. Take a column of marchers (or a stream of water) moving past a line beyond which they all

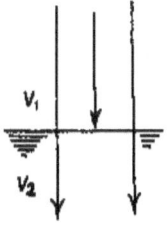

slow down. The fast marchers who follow will keep crowding on the slower ones, unless they avoid passing the line.

Actual intensity measurements confirmed that, similar to refractivity, reflectivity is a function of the angle of incidence; it increases as the latter increases. At grazing incidence, light is almost totally reflected. As Sommerfeld poetically put it, "This is the reason for the mirror image of the setting sun in a smooth sea; this image approaches the sun itself in intensity."[25] At the other extreme—at perpendicular incidence—little light is reflected. The more light remains on the side from whence it comes, the less transmitted, and the greater the refraction (we also remember that light is more refracted the slower it moves in the denser medium).

The relation of reflectivity to velocity is evident also when light strikes the interface from the side of the denser medium, in which case reflection is denoted internal. We have seen that Snell's formula applies here only up to a certain angle of incidence—Kepler's critical angle—where reflectivity is total and no light traverses the interface. The critical angle as determinant of internal reflectivity is a function of light's velocity in the medium (refractive index). When the rarer medium is, say, air, the critical angle in glass is smaller than in water, where light travels faster.[26] If the velocity of light in water is decreased (as may be done by cooling it) while the angle of incidence remains constant, a point is reached where total reflection starts.

In short, reflectivity here is a function of light's velocity in the medium; it increases as the latter decreases. Low velocity means that less light arrives at the interface per unit time; it is less concentrated, less intense.

Huygens already thought that "experience, moreover, teaches us that these two reflections [internal and external] are of nearly equal force, and that in different transparent bodies they are so much the stronger as the refraction of these bodies is the greater."[27] (The first part of this idea, however, had already been shown inaccurate by Kepler, for whereas there is an angle of total internal reflection, none exists for external reflection.)

Similarly, Newton said, "That in the passage of Light out of Glass into Air there is a Reflection as strong as in its passage out of Air into Glass, or rather a little stronger." Bouguer finally revealed by actual empirical measurements that total internal reflection was the strongest.[28]

When the violet part of a spectrum arrives at an interface from the denser side, it is totally reflected at an appreciably smaller angle of incidence than the red, which at this angle is still refracted out, i.e., the reflectivity of violet is higher than that of red. In Newton's words, "The Sun's Light consists of Rays differing in Reflexibility, and those Rays are more reflexible than others which are more refrangible."[29]

The fact that the violet's angle of deviation by refraction is greater than that of red indicated to us earlier that the velocity of violet in the denser medium was lower. And since its reflectivity is also greater than that of red, we conclude again that the slower the light, the higher its reflectivity. The more light is reflected at the interface, the less passes through, and the more deviated after transit, or refractivity is proportional to reflectivity.

In summary, we note that light's reflectivity at an interface is a function of the angle of incidence; it grows as the latter does. At the same time, reflectivity is also a function of light's velocity in the medium; it grows as the latter diminishes. Thus, we have two causes: increased angle of incidence and lowered velocity, which result in equivalent reflective effects. We may, therefore, logically suspect that an increased angle and lowered velocity have something in common.

5. Intensity

The term *intensity* denotes the amount of light and is applied to two distinct but related entities. One is the quantity of light emitted by its source—the luminous intensity or luminous flux—the other refers to the quantity of light on a receptive surface some distance from the source—the brightness, luminance, or illumination. It is the latter that concerns us here, presupposing that the luminous intensity and location of the source are known and invariable. Concerning luminous intensity and flux, Thomas Young objected to Newton's concept of light. "How happens it that, whether the projecting force is the slightest transmission of electricity, the friction of two pebbles, the lowest degree of visible ignition, the white wind of a heat furnace, or the intense heat of the sun itself, these wonderful corpuscles are always propelled with one uniform velocity? For if they differed in velocity, that difference ought to produce a different refraction."[30] Indeed they do, but this was not demonstrated until a few decades later by Friedrich Zöllner.

When a given quantity of light present on a small area is distributed over a larger one, the amount of light per unit area diminishes. Considering an obliquely transected beam, it is clear that its cross-section area increases as the angle of transection turns aside from the normal to the beam (figure). More precisely, the diminution of intensity F relates to the original intensity I as the cosine of incidence: $I'=I\cos\alpha$. This is known as the cosine law of surface brightness after Johann Heinrich Lambert (1728-1777).[31, 32]

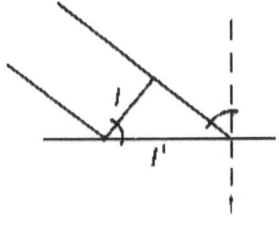

Trigonometrically, the cosine is the reciprocal of the sine, i.e., when it is at a minimum, the sine is at a maximum and vice versa. The cosine also means that as the angle of incidence grows in a simple manner, the intensity on the interface diminishes at an accelerated rate. Moreover, since area size relates to the *square* of the radius, and line I' is longer than I, the rate of intensity diminution attended to increased obliquity is actually that much greater: $(I)^2/(I')^2 < I/I'$.

In transit from a rarer to a denser medium, the total quantity of light in the transmitted refracted beam is reduced by the amount lost to reflection, and its quantity

27

per area is lessened by its obliquity to the incident beam (the angle of deviation); both diminutions increase with the angle of incidence. In addition, not only are the events *covariant* but the *rate of change* in all three factors—waning intensity, waxing reflectivity, and increased refractive deviation—grows progressively with the angle of incidence.

Since light manifests itself only in motion, it is necessary to include the element of time when dealing with its quantity. At any given period of time, the amount of light falling on a surface depends on its velocity; the slower the flow, the less per unit time strikes the surface. Intensity is, therefore, also defined as "the time average of the amount of light which crosses in unit time, a unit area perpendicular to the direction of the flow."[32]

For equal sources of light and equal areas of interface, the quantity of light incident per unit time from the side of the denser medium is less than that incident from the rarer one because (1) its velocity in the denser medium is lower; and (2) the total amount that traverses the interface is further reduced by reflection, which is also greater on the side of the denser medium. Finally, upon transmission, and for equal angles of incidence on either side, the light thus reduced deviates more when it arrives from the side of the denser medium than when it arrives from the rarer one. When the intensity is then further reduced by increased obliquity, a point is reached where light altogether ceases to cross the interface (critical angle).

The circumstances are also illustrated by the colors of a spectrum that in a denser medium travel at various velocities, the violet slower than the red. The intensity of violet on a given area per unit time is thus lower than that of red; its reflectivity (critical angle) is higher, and its refractive deviation upon transmission is greater.

Another element in the discussion of intensity is named after the father of photometry,[33] Pierre Bouguer (1698-1758),[34] of whom Priestley said, "No philosopher since the time of Sir Isaac Newton has given so much attention to the subject of light as M. Bouguer, and, next to those of that great philosopher, his labors seem to have been the most successful."[35] Bouguer's law states that the intensity diminishes exponentially with *thickness* of the medium and with its coefficient of absorption. It means that immediately after having passed an interface into a denser medium, the intensity of light continues to diminish.

In summary, the intensity of a given beam on an interface relates to the surface area and the velocity of light—it is lower the larger the cross-section area of the beam or the lower light's velocity or both.

A. The cross-section area is covariant (grows progressively) with the angle of incidence. Reflectivity—external and internal—is covariant with (1) the obliquity of incidence that determines the cross-section area; and (2) reflectivity also increases as the velocity in the denser medium is lower, hence reflectivity appears on both accounts to be a function of the intensity.

B. Refractivity is covariant (progressively increases) with (1) the obliquity of incidence that determines the cross-section area, (2) lower velocity in the denser medium (velocities ratio), and (3) reflectivity. Hence, refractivity appears a function of the intensity.

C. Dispersion is covariant with (1) the obliquity of incidence that determines the cross-section area, (2) with velocity in the denser medium, (3) with reflectivity, and (4) with refractivity. Hence dispersion appears a function of the intensity.

Tyco Brahe (1546-1601) and many others also believed long ago that refraction varied with the intensity of the light source. The sources were the sun, the planets, and the stars; and the refraction was measured by the degree of deviation produced by a prism. No such correlation was observed with this method. Earlier in the nineteenth century, Maxwell[36] and others concluded that refraction ought to correlate to the velocity of light, and they attempted to detect it by observing through telescopes the degree of deviation produced by a prism. Again, no such correlation was observed. The story of how this happened unfolds as we continue to examine the nature of light in the optokinematic part.

6. TEMPERATURE

The effect of temperature on refraction was examined in great detail by Armand Hippolyte Louis Fizeau (1819-1896), who utilized a complex apparatus to measure the phenomenon of interference.[37] By noting the change in the position of interference bands, formed by light transmitted through optical media at different temperatures, Fizeau concluded that refraction was directly proportional to temperature, i.e., the hotter the medium, the more refractive it was. The conclusion is more astounding since J. Albert Euler (1734-1800), son of the famous Leonard, had in 1762 already proven the opposite, namely that refraction was inversely proportional to temperature.[38] He arrived at this by measuring the change in the focal length of lenses interposed with water of different temperatures.

The actual state of affairs may be simply ascertained by a couple of **experiments.***

1. A narrow beam of light is obliquely transmitted through an aquarium about 20 cm wide filled with warm water. The position and width of the resultant image is marked on the opposite wall of the tank and on a screen several meters away. Ice cubes are then added to the water and the effect on the refracted light noted. After some minutes, it will be seen that the position of the image is moving in direction of the incident light, and its width seems to narrow, that is, the colder water at the bottom of the tank refracted light more and dispersed it less than the warm water.

2. Light is transmitted through a freezing cold prism resting on an electric hot plate. The width and position of the ensued spectrum are marked on a screen a few meters away. As the plate and the prism are slowly heated, one sees the spectrum widening and moving toward the prism's apex, i.e., the hot part of the prism refracted light less and dispersed it more than the cold prism (figure).

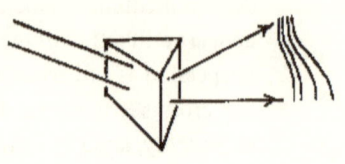

The effect of temperature variations on refraction appears to be opposite to that on dispersion, for whereas the angle of total refractive deviation grows with cooling, the distance between the red end of the spectrum and its violet end declines. It stands contrary to Newton's idea that refractivity and dispersion (refractivity of colors) were proportional, an idea first proven false in 1753 by John Dollond to whom we shall return when discussing the spectrum. Should this disparate effect of temperature be further confirmed, it could lead to construction of achromatic systems by temperature variations of one and the same material, similar as Dollond had done with different materials. The effect of different chemicals in lens material seem to have their counterpart in temperature variations, though the nature of this relationship between the chemical and the physical realm does not yet seem readily apparent.

---

*

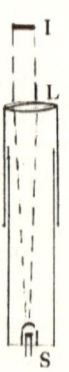

Unless otherwise noted, the light source in my experiments was furnished by a straight wire in a small 3 volts incandescent lamp (S) behind a collimating lens (L), forming a fairly linear image (I) manufactured by Bausch & Lomb Company under the name Copeland Streak Retinoscope. This seemed to correspond with Max Born's understanding, "Natural light, i.e., for light obtained from a body which is made to glow by raising its temperature" (see Born: Ref. 22 p. 44). Ordinary tap water and glass or plastic plates and prisms served as the denser medium in the rarer air. My prisms face measured 3 × 3 cm, and their refractive index was about 1.55.

According to the kinetic theory of thermal energy, heat is a manifestation of agitated motions of the medium's particles. The higher the temperature, the faster this motion. Light is similarly a motion; its intensity is related to its rate of motion. If one light is more refracted than another, it means that its loss of velocity by the medium is higher. Conversely, if a medium is more refractive at one temperature than another, its kinetic energy is lower; or the greater the gradient between the kinetic energy of light to that of the medium, the higher the refraction.

It is instructive to note in this connection that whereas at low temperature of optical media the linear propagation of light seems impeded (and in recent experiments was almost stopped completely), low temperature of electrically conductive media facilitates the flow of this current (superconduction). Refraction of gases also increases with increased pressure on them. We thus have three separate causes that result in the same effect: increased pressure, high refractive index, and low temperature all increase the deviation by refraction. One may therefore expect some relation between the causes. In our discussions, we presuppose that temperature and pressure remain constant and may, therefore, be ignored.

LOGICAL INTEGRATION

Six factors are basically involved in the encounter of light with an interface:

1. The angle of incidence upon the interface
2. The intensity of light on the interface
3. The velocities of light on either side of the interface
4. The reflection or reflectivity of light, i.e., the ease of its deviation by reflection
5. The refraction or refractivity of light, i.e., its ease of deviation by refraction
6. The dispersion or dilation of the beam

All six factors are variables in one and the same event that occurs in a very small area of space in a very short period of time. Up to a point, the sequential spatial steps are fairly clear: (1) light travels in one medium at a certain uniform velocity, (2) it arrives obliquely at an interface whence its intensity declines, and (3) it then proceeds with certain intensities and velocities in both media.

Despite light's great velocity, it is possible to also deduce a temporal sequence; a change in direction of light's flow toward the interface—the angle of incidence—occurs before it is incident on it, and this incidence—the arrival at the interface—occurs before light proceeds to travel by reflection and refraction. It therefore appears that events prior to arrival at the interface, and those on it, precede and lead to the events that follow. Other times the sequence is not readily apparent, for instance, does reflection occur before or after refraction or simultaneously with it? Perhaps, according to the different speeds, reflection in the rarer medium where light travels faster occurs prior to refraction into the denser one, whereas refraction into the rarer medium occurs before reflection in the denser medium.

It is quite usual and logically valid to assume a causal relationship between events that occur sequentially in place and time, particularly when space and time intervals are extremely small and no other factors are in sight. When our foot hits a rock and we fall right there and then, we logically conclude that hitting the rock caused the fall, particularly when nothing else is around to have caused it.

In the following examples of syllogistic deductions, Refr denotes refractivity; Refl denotes reflectivity; Int, intensity; Vel, velocity; Angl, angle of incidence; R, red; and Vi, violet. When one factor is *directly proportional* to another (=), it means that its magnitude varies in the same direction as the other, $x = f(y)$. When it is *inversely* proportional ($\neq$), it means its magnitude changes in the opposite sense than the other, $x = f(1/y)$. $\therefore$ denotes the logical *inference* from the premises. The two entries on the left are the *premises* induced empirically or by definition as described in detail in the previous pages. The entry on the right is the deduced conclusion, which in the physical realm requires further empirical confirmation. There is no reason why these relations may not be expressed mathematically by a suitably versed person.

Int.$\neq$ Angl;	Refl = Angl;	$\therefore$	Refl $\neq$ Int.
Int. $\neq$ Angl;	Refr = Angl;	$\therefore$	Refr $\neq$ Int.
Int. = Vel;	Refl $\neq$ Vel	$\therefore$	Refl $\neq$ Int.
Int. = Vel;	Refr $\neq$ Vel	$\therefore$	Refr $\neq$ Int.
Refl = Angl;	Refr = Angl;	$\therefore$	Refr = Refl.
Refl = Vel;	Refr = Vel	$\therefore$	Refr = Refl.
Refr = Refl;	Refl $\neq$ Int;	$\therefore$	Refr $\neq$ Int.
Refr Vi > Refr R;	Refr $\neq$ Vel;	$\therefore$	Vel Vi < Vel R.
Refr Vi > Refr R;	Refr $\neq$ Int;	$\therefore$	Int Vi < Int R.

Inferences and deductions, assuming that they are logically valid, may or may not be physically true. All reasoning is fraught with uncertainties, dubious applicabilities, and subtle fallacies where symbolic logic—such as mathematics—is further complicated at each step by the necessity to translate the symbol into its real physical meaning. Nonetheless, unless we wish to stumble blindly ahead, a map or paradigm helps. The final arrival place may turn out to have little to do with the accuracy of the map, and indeed may be reached by various routes, but some fairly valid line of reasoning helps, as exemplified by Columbus who relied on a faulty map to arrive at new knowledge that, in turn, helped improve the map.[39]

Reciprocity

The foundation of present knowledge in optics was laid in the Renaissance largely upon the medieval canon written by Abu-Ali Muhammad Al-Hasan Ibn Al-Haitham (al Basri), commonly known as Alhazen (c 965-1038). His Arabic treatise was translated into Latin by Vitello (1279), printed (1572) in seven books titled *Opticae thesaurus Alhazeni* and is generally acknowledged to have been the greatest advance since Ptolemy. Alhazen was the first to clearly express the idea that refracted light will follow the same routes whether incident on the interface from the rarer or the denser medium.

On the first page of his *Dioptrice*, Johannes Kepler (1611) designated Alhazen's idea as the third principal law of optics: "III. Axioma opticum. Eadem est refractio radiorum sive illi natura sua ingrediantur sive egrediantur, velut tales considerentur"—the deviation by refraction is the same whether the rays enter or exit, whichever way we consider it.[40] Isaac Newton may have read Kepler, though he didn't say so, for in his *Opticks*, he also designated the idea as the third axiom: "Ax III. If the refracted Ray be returned directly back to the Point of Incidence, it shall be refracted into the Line before described by the incident Ray."[41] Huygens said, "The refractions are reciprocal between the rays entering into a transparent body and those which are leaving it."[42] Robert Smith in his influential *Compleat system of opticks* in the eighteenth century included reflection when he stated the "third Law: If the reflected or refracted ray be returned directly back to the point of incidence, it shall be reflected or refracted into the same line before described by the incident ray."[43]

In the nineteenth century, the proposition did not change; John Tyndall enunciated, "A principle which underlies all optical phenomena—the principle of reversibility. In the case of refraction, for instance, when the ray passes obliquely from air into water, it is bent *towards* the perpendicular; when it passes from water to air, it is bent *from* the perpendicular, and accurately re verses its course"[44] (his emphasis). In the last century, Ernst Mach and others said of

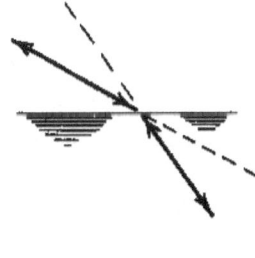

the reciprocal law, "A ray crossing the interface between two media, when reversed, retraced its original path."[45]

Inquiry into the circumstances of this or any other statement about nature usually follows Francis Bacon along two routes: one along the lines of reason, exploring the statement's logical validity, while the other seeks to verify its actual physical manifestations. Inasmuch as the law of reciprocity has already been stated in defined terms, we begin by exploring their meaning. The law contains terms that cause some difficulty by their ambiguity. What was meant by the *same*—light, ray, or direction—and *reversed* or returned. Light, constantly flowing like a river, can never be the same a second time, particularly after it was altered by its encounter with an interface, to be reflected and refracted into the second medium where it was additionally partially absorbed. Does *return* apply to the original incident light returned back from the interface or from some undefined distance from it in the denser medium? Does the returned light perhaps consist of another light source of equal strength and at equal distance from the interface, but on the opposite side?

When light that travels in air arrives obliquely at an interface, it is partially reflected and partially refracted, whereby the intensity of the refracted beam is lowered by the reflection and by its larger cross-section area of the oblique incidence. Having traveled in the denser medium any distance at all, it is immediately further reduced by absorption and scatter. When this light is then returned to the interface from the denser side, for instance by means of a mirror, it is there further reduced by internal reflection. When we recall that refraction is a function of intensity, it is hard to see why the finally returned light emerging into the rarer medium should follow its original path. It seems that since its intensity had diminished, it ought to be refracted out at a larger angle than that of the original angle of incidence into the medium.

Before Snell's formula was adapted when it was believed that the angles of incidence and refraction were at a constant ratio, it was reasonable to deduce that it did not matter from what side of the interface light came from. After Snell, this position was harder to defend because the direction of the refracted ray did not vary directly anymore but with the sine of the angle of incidence; for angles of incidence of equal magnitude, the *deviation* was greater when incidence was from the side of the denser medium, and furthermore, when light arrived from this side, it required a smaller incremental *change* of angle to effect a given deviation of the refracted ray than the change needed to effect the same deviation when arriving from the rarer medium. Refractivity is not the same whether the rays enter or exit, whichever way we consider it.

If, however, Snell's equation is taken as a complete representation of refraction and is also expressed in mathematical terms, it removes the incentive to explore refraction and its reciprocity. For there is hardly any reason to explore anything that is believed already fully understood or axiomatic, and mathematically, it matters little whether one writes $\sin\alpha = n \sin\beta$ or $\sin\beta = \sin\alpha/n$. Reciprocity here exists by definition.

Searching the literature, I did not find a single empirical measurement or practical demonstration of the reciprocity law or axiom. It is, of course, one thing to retrace a line on paper and another to duplicate an event in space and time. Inasmuch as the measure of refraction is determined by a number of variables, which are different for light that is incident on one side from that incident from the other, we expect the ensuing effects to differ accordingly. In particular (for equally strong light sources), the intensity on the side facing the denser medium is lower than on the opposite side, and the deviation on denser-to-rarer transit would be expected to be greater than in the opposite transit.

Experiments. I tried a practical demonstration by casting light onto the wall of an aquarium where the area of incidence is bordered with black tape; the refracted light in the water—made visible by fluorescein or dust—is then returned to the point of incidence by means of a silvered mirror. Unfortunately, the inclined mirror forms with the interface a prism that disperses the mirrored light into a spectrum so that one must decide which part shall exit, the red end or the violet. Upon exit, the light is again dispersed, and the final deviation depends on the color, which introduces an extraneous factor to complicate a decision. In fact, however, I could not eliminate dispersion even with a one-colored (monochromatic) *laser* light of a pointer.

The difficulty may be overcome when the refracted light in the water, instead of being reflected back and thus causing this annoying dispersion, may be made incident on another interface *parallel* to the first. The angle of incidence on this second face is the same as if the light was returned to the first one. In addition to a water tank, I used a thick plane-parallel transparent plate of high refractive index, or a large series of parallel plates.

I let light A (figure) fall *very* obliquely on the plate and observed the finally double

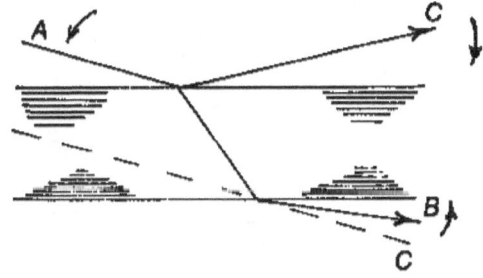

refracted light (B) on a screen a long distance away in order to more easily note the effects. In first incidence, the light is divided in two, one part (C) is reflected from the first interface while the other refracted and transmitted through both interfaces (B). As the angle of first incidence is slowly increased, the image of the reflected light (C) slowly approaches the reflecting interface as does the refracted image (B) approach the second interface, but at a faster rate. The angle which the reflected beam forms with the interface will be seen to be larger than the angle of the refracted beam. Furthermore, as the angle of incidence (A) nears grazing, the refracted light vanishes completely by dint of internal reflection while a reflected image remains. Hence, it appears that the path of light passing obliquely through a plane-parallel plate is deviated toward the plate.

The magnitude of this refraction depends, as all refraction does, on the angle of incidence, on the color which the light induces, and on its intensity on the second interface as determined by the thickness and absorptive power of the plate. The thicker and more absorptive the plate, the greater the total refractive deviation, and the smaller need be the angle of incidence that results in total reflection and nontransmission. Deviation by refraction implies that other phenomena associated with refraction, such as dispersion (see chapter on the spectrum and *nanospectra*), may be produced by oblique transmission through a plane-parallel plate.

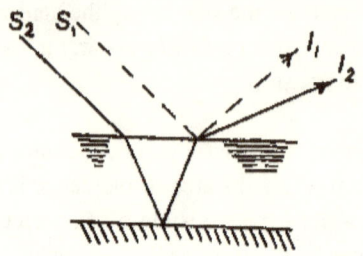

Take a case where two parallel rays (S_1 and S_2 in figure) fall upon a plate silvered on its second face (a glass mirror). The course of the refracted ray S_2 to and from the silvered face equals transit through double the thickness of the plate. The emerging ray (I_2) will deviate toward the plate at a smaller angle of inclination than the incident ray ($\beta < \alpha$). The reflected ray I_1 is assumed to reflect at about the same angle as its incidence. Therefore, on a distant screen, the two rays (*nanospectra*) will not coincide to form a single image composed of simple addition of their intensities, but will lie one next to the other to form a discontinuous image or fringes.[46,] The distance between the two images and their intensities vary with the distance of the screen from the plate since the rays are at an angle to one another—an easily verifiable fact.

A similar case prevails when two rays are incident at 90° upon both sides of a plate. The reflected ray S_1 (figure) arrives at I_1 and the refracted ray S_2 at I_2 some small distance away. The final composite image on the screen is, again, formed by rays that arrived to different localities with different intensities (fringes) (compare to Michelson's experiment).[47] The nanospectra from S_1 deviated at a different angle than from S_2 thus forming fringes, obviating the need for Young's and Fresnel's hypothesis of undulatory phase differences as their cause.

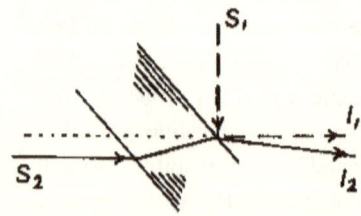

Let us now see how the refractive effect is manifested by interfaces that are not parallel but at an angle to one another, forming a prism.

Newton knew that in transit through a prism the total deviation had a certain minimum, for in his experiments, he so positioned the prisms that any rotation increased the deviation.[48] He then said that at this position, the angles of entry (first incidence) and exit (last refraction) were equal, i.e., the total transit was symmetrical

(AB in figure). His interpreter Robert Smith, however, missed the point by a wide margin when he reasoned that the angle of deviation always remained constant no matter how we rotate the prism because any increase on entry was compensated by a decrease on exit.[49]

Since the nineteenth century, the manner of transit through a prism that results in minimum deviation was rigorously deduced from Snell's law and the law of reciprocity, for instance by Brewster,[50] Born,[51] or Helmholtz. "It follows that in the symmetrical state we have the minimum deviation."[52] The conclusion that the deviation was at a minimum when the transit was symmetrical appealed to those who derived aesthetic satisfaction from seeing nature act economically and symmetrically—*harmonia mundis*—while those who had to face a sometimes cruel nature paid less regard to its aesthetics, recalling also perhaps that "nature will tell you a direct lie if she can."

The angle of minimum deviation at the prism's entry porta (G) is of course given at normal incidence when no deviation at all occurs. The same applies at the prism's exit porta (H). The total minimum deviation therefore occurs when the two normals are joined by the shortest line (G'H'), which geometrically must intersect them at equal angles. When one argues, as some have, that the light may be regarded as emanating from the center of this line (O) toward the prism's walls, then the angles of refraction will indeed be equal and the total deviation at a minimum (AB). On the other hand, when light arrives from one side is altered by the first interface, is then altered in transit through the prism, and again altered at the second interface, the proposition cannot simply be deduced mathematically but must be proven in fact.

According to our reasoning (which applies equally here as to parallel interfaces) when light passes through a prism in one direction—left to right or right to left, i.e., asymmetrically one way—the angle of deviation upon exit is larger than the angle of deviation upon entry (transit AB'). The angle of minimum deviation ensues when the angle of incidence on the prism's second interface is smaller than the angle of refraction at the first interface, which therefore results in asymmetrical transit (transit AC in figure). The true circumstances are the easier to ascertain, the larger the prism, and the higher its refractive index.

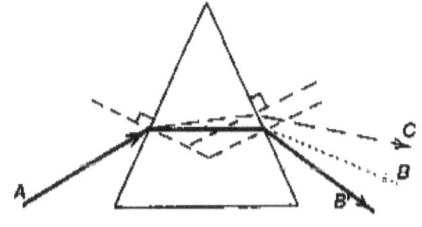

In as much as optical lenses are a composite of prisms, the circumstances of nonreciprocity apply here as well. In the figure, a ray from source S is refracted more than in the classical system where it would have gone to image I. In practice, however, the difference is imperceptible. In the "periscopic" lenses (meniscus) patented for spectacles by Dr. W. H. Wollaston,[53] the refraction on the back surface, where deviation and dispersion is higher, is diminished while the one in front increased, thus reducing their disturbing distortions.

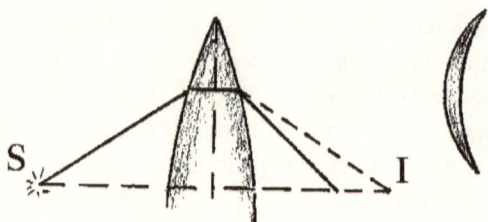

Concerning reciprocity, we may suspect that changing an axiomatic "principle which underlies all optical phenomena" would naturally lead to the need to consider anew our view of these phenomena. A thousand years after Alhazen, no reason is evident anywhere to compel us hold on to his ideas, particularly when the facts may be readily revealed with but some effort and little expense.

REFERENCES

1. Luce, A. A.: Logic. London. The English Univ., 1958, p. 27.
2. Hibben, J. G.: Logic. New York, Scribner, 1923, p. 258.
3. Euclid: Optica (in Opera Omnia, 7), Leipzig, Teubner, 1895.
4. Ptolemaeus, C: L'optique de Claude Ptolemae. Louvain, Univ. Catholique, 1956.
5. Crew, H.: The Photismi de Lumine of Maurolicus. New York, Macmillan, 1940, p. 49.
6. Kepler, J.: Astronomiae pars optica. Frankfurt, Marnium & Aubrisi, 1604, p. 111.
7. Kepler, J.: op. cit. Ref. 6, p. 115.
8. Mach, E.: The Principles of Physical Optics. New York, Dover, 1926, p. 32.
9. Alhazen: Opticae thesaurus. New York, Johnson Reprint, 1972.
10. Mach, E.: op. cit. Ref. 8.
11. Descartes, R.: La Dioptrique (1637), in: Magie, W. F.: A Source Book in Physics, Cambridge, Harvard Univ., 1965, p. 272.
12. Larmor J.: Aether and Matter. Cambridge Univ. Press; 1900: 310.
13. Fermat, P. In: Magie WF. A Source Book in Physics. Op.cit. Ref.11; p 278.
14. Feynman R. P.:QED; Princton Univ. Press, ; 1985: 51-52
15. Chales, C. F. M. de: Cursus seu mundus mathematicus. Lyon, Exofficina Anissoniana, 1674, chap. 23, p. 648 (Also: Editio altera, vol. 3, 1690).
16. Huygens, Ch.: Traite de la lumiere. Leiden, P. Vander, 1690. Transl. S. P. Thompson: Treatise on Light, New York, Dover, 1962, p. 3.
17. Baker, B. B., Copson, E. T.: Huygen's Principle. Oxford Univ., 1939.
18. Sabra, A. J.: Theories of Light. London, Oldbourne, 1967, p.216.
19. Foucault, L.: Method generale pour mesurer la vitesse de la lumiere dans l'air et les milieux transparents. *Comp. Rend.* 30;551, 1850.
20. Newton, I.: New Theory about Light and Colors. *Phil, transact.* 80;3085, 1672.
21. Newton I.: Opticks. New York, Dover, 1952. p. 5-6.
22. Born, M. Wolf, E: Principles of Optics. 3rd ed., New York, Pergamon,1965, p. 13.
23. Helmholtz, H. F.: Handbuch der Physiologischen Optik 2nd ed., Hamburg, Voss, 1896, p. 49.
24. Ditchburn, R. W.: Light. London, Academic Press, 1976, p. 517.
25. Sommerfeld, A.: Optics. New York, Academic Press, 1950, p. 21.
26. Mach, E.: op. cit. Ref. 8, p. 121.
27. Huygens, Ch.: op. cit. Ref. 14, p. 42.
28. Bouguer, P.: Traite d'optique. Paris, Guerin & Delatour, 1760. Also: Optical Treatise on the Gradation of Light. Univ. of Toronto, 1961.
29. Newton, L: op. cit. Ref. 19, p. 63.
30. Young, T.: A Course of Lectures on Natural Philosophy, vol. 2, London, Johnson, 1807, p. 79.

31. Lambert, J. H.: Photometria, etc., Klett, 1760. Also: Photometric Leipzig, Englemann, 1892.

32. Born, M., Wolf, E.: op. cit. Ref. 20, p. 182.

33. Bouguer, in: Diet, of Sc. Biog. 2, New York, Scribner, 1973, p. 343.

34. Bouguer, P.: Essai d'optique sur la gradation de la lumiere. Paris, Jombert, 1729.

35. Priestley, J.: The History and Present State of Discoveries relating to Vision, Light and Colours. London, J. Johnson, 1772, p. 405.

36. Maxwell, J. C.: On the Influence of the Motion of the heavenly Bodies on the Index of Refraction. *Phil, transact.* 158; 532, 1868.

37. Fizeau, H.: Sur les modifications que subit da vitesse de la lumiere dans le verre et plusieurs autres corps sons Finfluence de la chaleur. *Ann. de chem. et de Phys.* 66; 429, 1862. Also in: *Ann. d. Phys.* 119; 87, 1863.

38. Hoppe, E.: Geschichte der Optick. Wiesbaden, Saendig, 1967, p. 66.

39. Kuhn, T. S.: The Structure of Scientific Revolutions, 2nd ed., Univ. of Chicago, 1970, p. 108.

40. Kepler, J.: Dioptrice. Augsburg, D. Franke, 1611.

41. Newton, L: op. cit. Ref. 21, p. 5.

42. Huygens, Ch.: op. cit. Ref. 16, p. 35.

43. Smith, R.: A compleat system of Opticks. Cambridge. C. Crownfield, 1738, p. 3.

44. Tyndall, J.: Six lectures on Light. 4th ed., London, Long mans, 1885, p. 17.

45. Mach, E.: op. cit. Ref. 8, pp. 29, 31, 39.

46. Born, M., Wolf, E.: op. cit. Ref. 20, pp. 61, 282.

47. Wood, R. W.: Physical Optics. 3rd ed., New York, Dover, 1967, p. 292.

48. Newton, I.: op. cit. Ref. 21. p. 28.

49. Smith, R.: op. cit. Ref. 43, p. 12.

50. Brewster, D.: A Treatise on Optics. Philadelphia, Lea & Blanchard, 1841, p. 32, and app. p. 28.

51. Born, M.: op. cit. Ref. 22, p. 178

52. Helmholtz, H. F.: op. cit. Ref. 23, p. 290.

53. Wollaston W. H.: Experiments showing the advantage of periscopic spectacles. *Phil. Mag.* 18; 1804: 165-66.

Colors

When light is more or less abruptly deviated from its straight course by encounter with matter—as it is in refraction at an interface—one of the ensuing effects is the perception of colors. Since refraction is a compound phenomenon with a number of variable parameters, the idea lies near that the concurrent phenomenon of color perception may similarly be affected by these parameters. But before entering into the discussion of color, the term must be defined, for it is this definition that seemingly posed an obstacle on the road to clarity.

The study of color has been described as a "difficult and fatiguing subject" that produced "an exceedingly abundant volume of material, so abundant as to make a good insight into it and a good survey of it extraordinary difficult."[1] Or else, "in no branch of knowledge has a theoretical edifice so large been built on a factual basis so small as in the subject of color vision."[2] Nobel laureate Max Born resigned to say, "The actual connection between color and frequency is very involved and will not be studied in this book."[3] Since, nevertheless, color is a pleasant and ubiquitous feature in optical phenomena, we may at least ask ourselves why is its study so terribly involved as to be avoided altogether in the most reputable books on optics and wherein lies this complexity?

Colors are infinite in number and are sensually accepted in equally numerous modalities; any one color conceived as beautiful by one observer appears ugly to his fellow, attractive to one is repulsive to another. This variable and unpredictable subjective *sensation* suggested that perhaps color itself was a subjective phenomenon that therefore varied with the observer's mood or mental constitution and, hence, could not be quantified in order to be studied in definite physical terms. Aristotle, and to some extent Goethe,[4] believed that colors were a mixture of black and white where equal parts of each produced green. Green was therefore the most pleasant and salutary color, and since, the Renaissance hospital walls and then surgeons' gowns were colored green.

Newton attached (1671) to each color the *physical* characteristic of refractivity manifested by a certain *geometrical* angle: "To the same degree of refrangibility ever belongs the same color, and to the same color ever belongs the same degree of refrangibility." Young, a century later, interpreted it in quantities of *length*, the length of waves or their *frequency*: "The sensation of different colours depends on the different frequency of vibrations excited by the light in the Retina." Yet there remained a number of phenomena that could not be understood in those terms, most notably but not alone, the production of one color by the combination of two, three, or any number of others. Allowing Newton's and Young's views, it then became necessary to assume the existence of as yet unknown mechanisms in the perceptive organs—the eye or brain—that apparently produced the other puzzling color phenomena, and consequently, a large number of theories of color perception were born, the Young-Helmholtz three-color theory being most popular.[5] But nevertheless according to Land, "The eye has recently been found to be an instrument of wonderful and unsuspected versatility. It can perceive full color in images, which according to classical theories, should be monochromatic."[6]

Thus, whereas the ordinary endeavor of research was to explicate physiology on physical and chemical grounds, phenomena of color had the unique distinction of physical events being elucidated, as far as it went, on physiological grounds. Having relegated color to the realm of physiology and cognitive neuroscience, the omission of its discussion in physics books was seen quite just, which oddly did not inhibit their authors from dealing on every page with one-colored (monochromatic) light.[3]

The subject of epistemology deals with how we know that we know, and since antiquity, one wondered about the accuracy of the real external world as perceived by our senses. Information about the vast majority of our experiences is visual—mediated through the eye to the brain—and we have come to rely on the eye, if only for lack of comparably accurate alternatives. Measuring rods and dials, manometers and thermometers, spectrographs and photographs—all are read visually with little doubt about the accuracy of the perception. Actually, light as a sensation manifests itself also as heat—ordinarily mediated through the skin to the brain—but no optics or other exact sciences are likely to be built upon these dermal data. The change in light's course by refraction is, after all, also a visual experience that is accepted as physical; when this refraction is attended by colors, there was no good reason to reject them as having no physical basis. But these questions remain: In what physical terms shall we define colors? What values of magnitude describe them?

Until we have found the physical *cause* of color, we cannot define it but in terms of its visual *effect*. Therefore, it would seem proper to term light that produces, say, the effect of red "red inducing." In 1671, Newton wrote, "Light is a confused aggregate of Rays indued with all sorts of Colours," but in 1704, he put it somewhat differently.[7]

> The homogeneal Light and Rays which appear red, or rather make objects appear so, I call Rubrifick or Red-making . . . And if at any time I speak of Light and Rays as coloured or endued with Colours, I would be understood to speak not philosophically and properly, but grossly, and accordingly to such Conceptions as vulgar People in seeing all these Experiments would be apt to frame. For the Rays

to speak properly are not coloured. In them there is nothing else than a certain Power and Disposition to stir up a Sensation of this or that Colour.

Regrettably, because of the immediate identification of visual effects with physical causes, Newton had great difficulty following his restricted definition and terminology, but rather spoke throughout of "the Colours of homogeneal Lights," "lights which differ in Colour," and "Colours are the qualities of Light." He also noted that the sensation produced by one color-inducing light did not seem to change further by any then-available means—"so far as I could judge by my Eye."[8] This "unchangeableness of Colour" led him to conclude that color induction was not an event caused by a change in light through refraction but was a property (its refrangibility), a connate property specific to each light.

Newton had shown that Snell's formula was insufficient to explain refraction, seeing that refraction was not merely a function of the angles of incidence and refraction but also varied with some other effect—an effect that elicited the sensation of color. Logically, what was hitherto a useful translation of a physical event (refraction) into geometrical terms (angles) was enlarged to its exposition in terms of its physiological effects (color sensations) that were themselves defined in terms of the original physical event (refractivity).

The second fundamental contribution of Newton followed from his first. "But the most surprising and wonderful composition was that of Whiteness. 'Tis ever compounded, and to its composition are requisite all the aforesaid Colours, mixed in a due proportion." Or "Whiteness is a dissimilar mixture of all Colours, and Light is a mixture of Rays endued with all those Colours." The refracting prism acted simply as a sieve that disentangled the passing colored light globules according to their different sizes or other characteristics.[9]

Atomistic mode of reasoning such as Newton's suited chemistry well and was applicable to light when light was considered a composite of atoms of various size; beyond this, it was unfortunate. The concept of white light as being a loose mixture of single colored lights led to the conclusion that all color-inducing lights must have the same velocity (in air or vacuum). "If the velocities for different rays were different in vacuum, the aberration of stars (which is inversely as the velocity) would be different for different colours, and every star would appear as a spectrum whose length would be parallel to the direction of the earth's motion. We know of no reason to think that this is true."[10] Like a herd of horses let loose, all must trod at the same speed, otherwise one horse will reach the finish line before the others, i.e., white light will change color from, say, red to yellow to violet, which does not conform to reality. [11]

Following Young's conception of light in terms of length of waves (or frequency), Newton's idea concerning color was expressed in those terms. "The colors of light consist in the different frequency of the vibrations of the luminous ether: the opinion is strongly confirmed by the analogy between colours of a thin plate and the sounds of a series of organ pipes."[12] The length of undulation in air of the red-inducing light was given as 0.000266 parts of an inch or, later, about 700 nanometers. Each color-inducing light had its specific connate property—this time the length of the wave, coupled to refractivity—and white-inducing light was a mixture, superposition, of the waves of color-inducing lights (white = many colors, in Greek: polychromatic). The color sensations induced by dispersion that accompanied refraction were not a new event due to a change in the nature of the refracted light, but rather were an inherent quality (the wave-length) that was only unmasked by oblique transit through the prism. As in a modern textbook,

> if a ray of polychromatic light is incident upon a refracting surface, it is split into a set of rays, each of which is associated with a different wavelength. In traversing an optical system, light of different wavelengths will therefore, after the first refraction, follow slightly different paths.[13]

Early in the nineteenth century, it became evident that each and every receptive organ had its very own specific response to stimulation; no matter what the physical stimulus, the receptor always produced one and the same sensation. This is known as the law of specific nerve energies after Johannes Mueller (1801-1858).[14, 15] No matter what the stimulus to the eye—mechanical pressure or tension, light, X-rays, or electricity—the eye responds only be inducing the sensation of light.[16]

Newton pressed on his eye with his finger and a bodkin (point *a* in figure) and saw colored lights (area c). Since colors are perceived by light and also by pressure, it suggested that light exerts pressure.[17]

The final channel of information being limited in this manner, it would be quite inadequate, and indeed deceptive, to study mechanical forces, electrical currents, or X-rays and directly deduce their laws from their unmediated actions onto the eye. Similarly, the mere statement that a particular color had been sensed does not necessarily mean that a particular physical cause brought it about; it may be so, but the evidence is insufficient because several physical variables in the stimulus may yet result in only one sensory response. And therefore, if we wish to understand the power of light to induce colors by refraction, we must explore the physical nature of this stimulating event. We may seek physical determinants of color sensations by quantitatively varying physical parameters in the event of refraction that produced them and note the ensuing effect, if any. We may then conclude that these physical entities at given quantities are, or are not, the cause of the effect. At this point, however, the exact quantitative data are of secondary importance because we must first establish that the measured entity is indeed of causative interest.

REFERENCES

1. Halbertsma, K. T. A.: A history of the theory of Colours. Amsterdam, Swets & Zeitlinger, 1949, p. 5.
2. Duke-Elder S (1973): System of Ophthalmology. St Louis, CV Mosby, 4: 435.
3. Born, M.: op. cit. *Principles of Optics*, p. 11.
4. Goethe, J. W.: Theory of Colours. London, F. Cass, 1967.
5. Linksz, A.: An Essay on Color Vision. New York, Grune & Stratton, 1964.
6. Land, E. H.: Experiments in Color Vision. *Sc. Am.* 5; 84, 1959.
7. Newton, I.: *Opticks*, p. 124.
8. Newton, I.: *Opticks*, p. 73.
9. Korye, A.: Newtonian Studies. Univ. of Chicago, 1968, p. 8.
10. Airy, G. B.: Undulatory Theory of Optics. London, McMillen, 1877, p. 35.
11. Einstein A. Relativity. 1961; New York, Crown: p 17.
12. Young, T.: A Course of Lectures on Natural Philosophy, vol. 2, London, Johnson, 1807, p. 543.
13. Born, M.: op. cit. p. 174.
14. Müller, J.: Handbuch der Physiologie des Menschen, 3rd ed., Coblenz, Holscher, 1838, p. 780.
15. Singer, C, Underwood, E. A.: A Short History of Medicine, 2nd ed., New York, Oxford Univ., 1962, p. 287.
16. Adler, F. H.: Physiology of the Eye, 3rd ed., St. Louis, C. V. Mosby, 1959: p 529.
17. McGuire JE, Tammy M. Certain Philosophical Questions. Cambridge Univ. Press. 1983: 481

The Spectrum

The term *spectrum*, as introduced by Newton, usually denotes a band of color-inducing lights in specific order, the red located at one end, gradually translating into orange, then yellow, green, blue, and finally violet at the other end. It manifests itself naturally in the rainbow. When a spectrum is formed by refraction of ordinary white-inducing light, the violet end is always the most deviated toward the denser medium and the red end, the least deviated. By refraction through a prism, the violet is nearest the base and the red nearest the apex.

Dispersion and the concomitant perception of spectral colors occur when light is refracted in either direction, from the rarer into the denser medium and in reverse. This

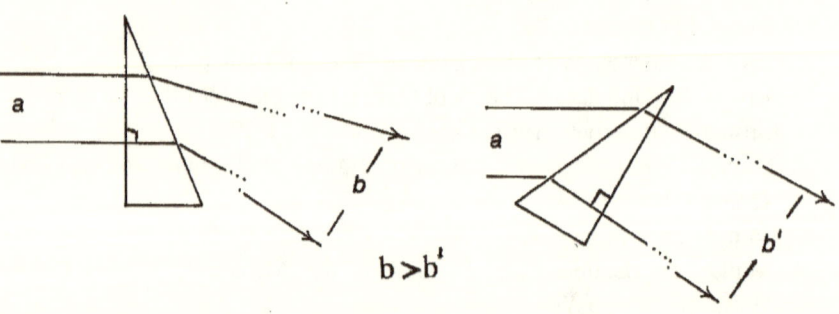

is evident when we cast light perpendicular to a prism's first interface where it is not refracted and then observe the spectrum on a screen beyond the second refracting interface (figure, left). The normality of incidence may be confirmed by noting that the reflected light off this first interface is returned in the direction of incidence. The prism is then rotated clockwise until refraction occurs at the first interface alone, permitting the refracted beam to leave the second face as accurately as possible, perpendicular and unrefracted. We again observe a dilated spectrum. Because of the limits of reciprocity, the normality to this second interface may be attempted by noting that the reflection is returned in direction of the incident beam (the angle of incidence here leads easily to the refractive index of the prism because the angle of refraction equals the angle of the prism).

We have noted in the previous section that refractivity—the ease of deviation by refraction—was greater in transit from the denser medium into the rarer one (b) than in reverse (b'). Similarly, the dilation of the beam by refraction (its *dispersion*) is greater in transit from the denser medium than it is from the rarer one. No reciprocity exists, and what was done in transit in one direction cannot be undone by reversing direction.

Experiment. I took a prism of 35° made of material with a refractive index of about 1.5 and a width of incident beam of about 10 mm (a, in figure.) From Snell's equation (or by approximate measurement), I found that the angle of deviation by refraction is about 25° in both positions of the prism. Trigonometrically (or by measurement), I found that the width of the beam immediately after refraction in the first position is about 6.1 mm (left in figure). When we then measure the width of the spectrum, some 4 meters beyond the prism, it is about 40 mm, i.e., the refracted beam dilated (dispersed) 33.9 mm, from 6.1 to 40 mm. In the second position (right in figure), the width of the beam near the prism is found to be about 16.4 mm, and 4 meters away—30 mm; it dilated 13.6 mm, from 16.4 to 30, much less than in the first position (b'< b).

Depending on the direction of light's transit through the refracting interface, from the medium where it is fast to the one where it is slower or in reverse, the position of colors in the spectrum varies; the distance between them—say between red and blue—is greater in denser-to-rarer transit than in the opposite direction. If we take the position of one color as a reference point or the average of all colors combined, it is evident that the angle of refraction that produces one and the same color sensation, say blue, varies; the angle is larger in denser-to-rarer than in rarer-to-denser transit. Or in general, the angle of refraction of one and the same color-inducing light is not constant; it is greater on transit from the denser medium than in reverse. Contrary to Newton, it seems, therefore, that the sensation of colors and the refractivity of lights that induced them were not related.

A full spectrum is not visible immediately behind a prism in either position. Instead, we see a white center fringed toward the prism's apex with colors from yellow to orange to red—the warm colors—and toward the base, the colors from yellow-green to blue to violet—the cold colors. Before a full continuous spectrum ensues, the screen must be moved farther from the prism when the refraction is from the rarer medium into the denser one than when refraction is in the opposite direction (b versus b'). This central whiteness right behind the prism caused Goethe to believe that the sensation of color had nothing to do with the physical event of refraction. "But how astounded was I when the white wall seen through the prism remained white as before."

47

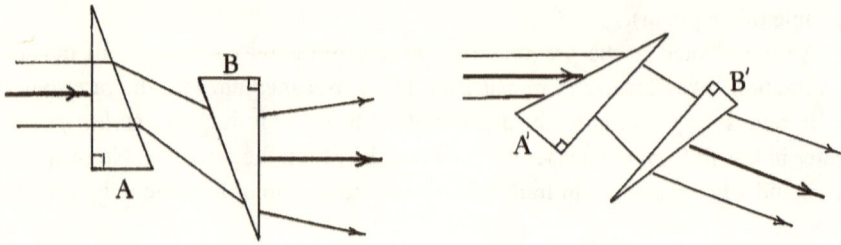

In both positions of the prism, the average angle of refractive deviation is about the same, that is, the general direction of the dispersed beams, or their yellow center, is the same. However, since the dispersion varies according to whether light arrives from the denser or the rarer side, it is possible to combine interfaces of only two media, such as air and one kind of glass, so that dispersion ensues without refraction or where refraction occurs with but little dispersion. On the left in the figure above, light is refracted and dispersed by the second face of the first prism (A). The deviation by refraction may be counteracted by the first face of another *identical* prism positioned upside down (B), but the dispersion remains. On the other hand, the dispersion by the first refracting face of the first prism (A' right in figure) is small and may be counteracted by the second face of a *weaker* prism (B') that is, however, insufficient to cancel the refraction. In this manner, one obtains achromatic refraction. It is not truly colorless because refraction and dispersion (refractive and dispersive coefficients), though generally correlated as Newton believed, do not correspond directly and exactly, being that the two are different events.

To obtain achromatism in practice, one does not employ prisms (or lenses) of the same material, as we did, but prisms of differing materials. Similar as refractivity relates to the nature of the material, its refractive index, so does the degree of dispersion vary for differing materials. For equal magnitudes of refraction, the position of the colors in the spectrum, and hence the distances between them, varies from medium to medium, a phenomenon termed irrationality of dispersion.[1] Light's behavior was termed irrational because the accepted theory was deemed rational. According to Newton, each color was tied to a specific refrangibility, yet here the same colors are differently refracted. According to Huygens, refraction was tied to the wavelength to which Young later ascribed the power to induce color, yet here are the same waves variously refracted.

There are few facts in the history of science more singular than that Newton should have believed that all bodies, when shaped into prisms, produced prismatic spectra of equal length, or separated the red and violet rays to equal distances when the mean refraction of the middle ray of the spectrum, was the same . . . And when, under the influence of this blind conviction, he pronounced the improvement of the refracting telescope to be desperate, he checked for a long time the progress of this branch of science, and furnished to future philosophers a lesson which cannot be too deeply studied.[2]

The variable refractivity of a light that induces the same color was first noted by John Dollond (1706-1761) who used this property to rid optical instruments of annoying chromatic aberration (dispersion).[3, 4] Fairly equal separation of colors is obtainable by unequal refraction in two different media. In the words of Dr. Young's friend Dr. Wollaston, "But it may happen that two media, which refract unequally at the same incidence, may disperse equally at that incidence."[5] When the two media (prisms or lenses) are then combined in opposite directions a residual refraction remains, but the dispersion is largely cancelled. On the other hand, equal refraction by two different media results in unequal dispersion, so that refractions in opposite directions cancel the deviations without eliminating dispersion.

This last scheme is employed in G.B. Amici's (1786-1863) direct vision prisms or in the Pellin-Broca constant-deviation spectroscope.[6] The fact that rays exit an ordinary prism at various angles (dispersion) made detailed study of the spectrum cumbersome by the need to focus the viewing telescope to different depths. Combining prisms of different materials cancelled the angular refraction and facilitated the task.[7] We shall touch on this method again later when dealing with absorption lines and the big bang.

As to the cause of the difference in the dispersion from the two sides of the interface, we note that in the denser medium (1) the velocity of light is lower than in the rarer one, (2) its intensity when it reaches the refracting interface is lower by passage through the more absorptive and internally reflective medium, and, above all, (3) the time difference between the arrival of one side of the incident beam to the interface and the arrival of the other side, the gradient of arrival time, is greater. On the other hand, light's density is somewhat higher on the side of the denser medium because the area is smaller, for equal angles of deviation the interface is less inclined to the incident light here than in the rarer medium. The irrationality of dispersion resides in the capacity of different media to scatter and absorb light (diminish its intensity) according to their constitution.

Since Newton, it was said that all spectral colors were contained in the incident white-inducing light. When white is dispersed by a prism into a spectrum, the red is located uppermost toward the prism's apex (its light being least refracted) and the violet is located nearest the base. The question then arises: where in the white light are the color-inducing lights located?

Newton illustrated the transit of rays through a prism as in the adjacent figure; the red-inducing ray issuing from the uppermost part of the incident beam, and so forth, down

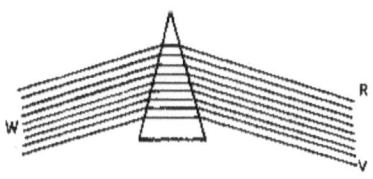

to the violet. But if all the red-inducing light, which forms the prismatic red image (R), is located at the top of the incident white beam and all the violet at the bottom, it is puzzling why this incident beam does not induce these colors at these locations without refraction—why does the sun or the moon not appear red on top and violet on the bottom?

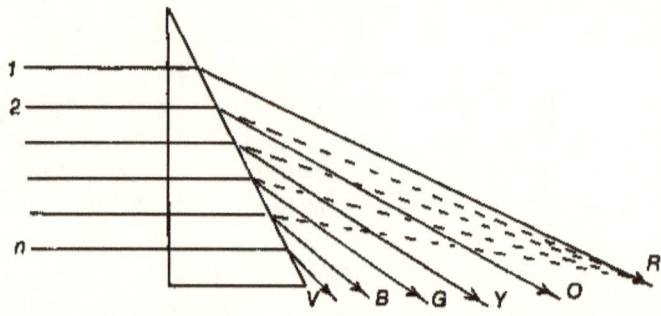

Or else (figure), if every part of the incident white beam (1, 2-n) contains elements of each color, say, red-inducing light, then all these red-inducing elements must be refracted at different angles (dotted lines) in order to reach their top location in the spectrum; namely, the red-inducing light in incident portion 1 more refracted than that in incident portion n.

Experiment. To gain insight into the actual circumstances, I proceeded to analyze the spectrum, literally loosen it into its separate ingredients. (Compare to Newton's experiment 3 in book 1, part 2 of the *Opticks*.) To this end, I cast obliquely a beam of white light upon a prism to create a broad spectrum on a wall some 4-5 meters away (R-V). To Newton, the red color in this spectrum was homogeneal, unchangeable, and not subject to further separation by refraction. It answered to its own very specific magnitude of refraction, which in case of red was less than that of the other colors. In the same vein, Thomas Young thought that this red had its own very specific wavelength, say, in the range of 650 to 700 nanometers, immutable and inseparable and corresponding to its own magnitude of refractivity. All the other colors were similarly homogeneal (monochromatic) and had their own wavelength down to about 400 nm for the violet.

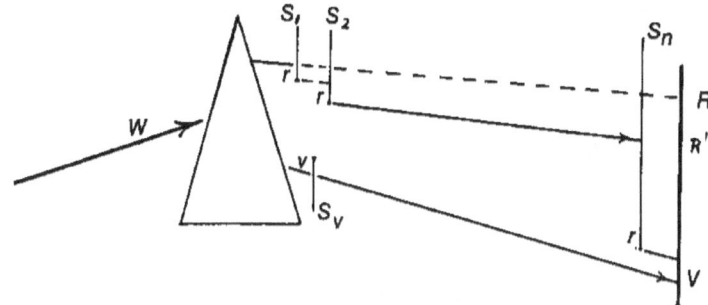

As I kept my eyes on the spectrum, I introduced a small screen into the refracted beam about 4-5 cm from the prism and from the side of the prism's apex (S_1), which screen intercepted the red color, with some orange perhaps for good measure. I expected the red of spectrum R-V on the wall to disappear so that the spectrum on top would begin with orange or yellow. But this did not happen; the red was still there though the spectrum narrowed (R'v). I then introduced another screen (S_2) from this side and about 4 cm farther down the beam in order to intercept the new red. Again, the red of the spectrum on the wall remained intact. I continued in this manner to place n number of screens into the refracted light, and all intercept only red until the final spectrum R-V had so narrowed as to be barely perceptible. I have intercepted on screens S_1 to S_n about all incident white light which was dispersed by the prism, and it was all red! Furthermore, since the spectrum was, as usual, wider than the incident light, the red on screen S_1 was obviously less refracted than the red on screen S_2 and certainly less refracted than the red on screen S_n.

I then repeated the experiment by successively introducing screens S_1 to S_n, but this time from the side of the prism's base. All these screens were now colored violet-blue—incident white turned almost completely to violet. And the violet on screen S_1 was obviously more refracted than the violet on S_n. In addition, the red on screen S_n, which was introduced in the first place from the apical side, was evidently more refracted than the violet on screen S_n introduced from the basal side. By placing screens into the refracted and dispersed beam, near the prism in the beam's midst or from either side to varying distance, I was able to change any color of the spectrum on the wall to any other color, diametrically contrary to Newton and Young.

Returning to our original prism and spectrum, I have introduced into the refracted light several successive screens S_1 to S_n from the apical side and noted that the spectrum had narrowed in such a manner that where we originally perceived green. For instance, I now saw red, and I also saw red on all the screens, including the farthest from the prism. I now placed an additional screen S_v near the prism, but from its basal side, in order to intercept the violet end of the beam emerging from the prism. We may expect to thereby extinguish the violet end of the spectrum on the wall, which did not disappear by introducing screens from the apical side. We discover, however, that this last screen S_v near the prism received violet-inducing light, instead of erasing the violet on the wall, extinguished the red on screen S_n on the opposite side of the beam. We intercepted violet and extinguished red!

How are these seemingly puzzling phenomena to be interpreted? I believe the task is eased when we regard the final spectrum on the wall as composed of—being an integral of—myriad small spectra, an atomistic approach. All these microspectra, or *nanospectra* (Latin *nanus*, dwarf), emanating from the prism are variously refracted, the top one the least, the bottom one the most. Three are pictured in the figure where the continuous lines stand for red and the interrupted ones for violet.

As screen S_1 was introduced, it received the red end of the uppermost nanospectrum (R in figure), allowing the red of the next lower nanospectrum to manifest itself on the distant wall (r'), and so forth to screen S_n. When screen S_v was introduced from the bottom, it received some violet, but this was replaced by the violet from the next higher nanospectrum. However, the red (r'') on screen S_n could not be replaced and became extinguished.

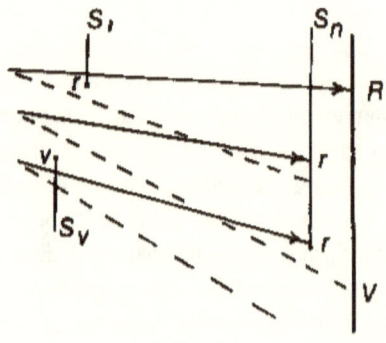

The color-inducing lights that form the top nanospectrum do not correspond to the lights forming the lowest one, having been produced by less refracted lights. Recombining the lights dispersed in the top nanospectrum results in a differently intense white than when reuniting (as by a convex lens) the color-inducing ingredients of the lowest nanospectrum. Similarly, substituting part of one nanospectrum, say the top half, with the corresponding part from another nanospectrum results in a complete spectrum; but one that does not possess an intensity value equivalent to the original. We shall return to the matter when touching on periodicity and interference.

Experiments. Independence of color from refractivity is demonstrable by various other means, for instance by modifying Newton's experiment 5 in book 1, part 1 of his *Opticks*. I cast an oblong beam of light perpendicularly to a prism's face and then observed the continuous dilated spectrum on a screen some distance away (RYV). Two black bands (electrician's tape) were then glued to the prism's face (A, B), in order to divide the incident beam into three (figure). On a screen (S) a short distance from the prism, three complete but separate spectra ensued that *farther away combined to form one continuous spectrum*, red on top violet on the bottom (RYV), similar to the

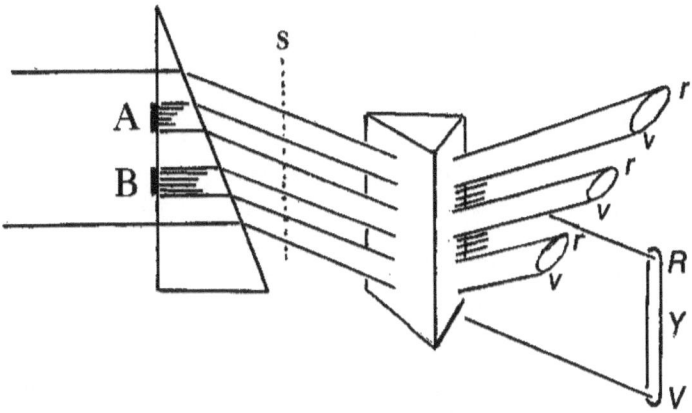

spectrum that ensued without separating the incident beam at A and B. The top spectrum near the prism corresponds to the top portion of the distant spectrum (red) and was least refracted; the lowermost of the three spectra was most refracted, and each individual color of the top spectrum was less refracted than the corresponding color in the other two.

I then placed a second prism in the three refracted lights a centimeter or so behind the first prism and perpendicular to it, whereby the refracted beams were refracted a second time but away from the plane of the page. On a screen parallel to the plane of the page, three full spectra ensued (rv). All the colors of the top spectrum that were least refracted by the first prism are also least refracted by the second one and are farthest removed from the base of the second prism. On the other end, the colors forming the lowest spectrum are nearest to the base, whereby the red, for instance, of this spectrum is obviously more refracted than the violet of the top spectrum.

The transmutation of spectral colors became evident also in the following manner. I cast a slightly convergent beam upon a prism and noted the position of the focal point where the rays intersect (F in figure). When a screen S_1 was placed in the refracted beam near the prism, a full spectrum was clearly visible in the usual order of colors, i.e., red at the top (RV). When the screen is moved beyond the focal point, the *order of colors does not change*—even though the rays have crossed, the red is still near the apex.

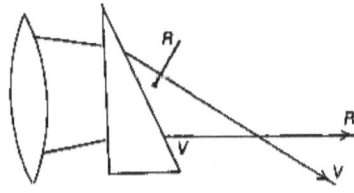

Leaving the screen in this position, I introduced a second screen immediately behind the prism from its apical side and, on it, saw the intercepted red (R). I discovered that this screen erased the violet in the spectrum formed after intersection. Similarly, a screen receiving the violet near the prism on its basal side (V) erased the red in the final spectrum. The colors in the spectrum prior to intersection answer exactly to the

53

opposite colors in the second spectrum. When red-inducing light is removed from the first, violet-inducing light disappears from the second.

The perceptual and conceptual difficulties underlying the crux of the matter are perhaps most simply illustrated by the following situation: An observer (O) is in a dark room with two holes covered with semitransparent paper or glass. An assistant outside the room casts on the wall a prismatic spectrum with its violet portion on the upper aperture (V in figure). He is asked by the man inside to transform this violet into red by changing the intensity, and probably the velocity, of the refracted light. To this end, the assistant rotates the prism a little, thereby changing the angle of incidence with the physical parameters attended to this change; and indeed the observer in the room perceives red (r) in the same place where he previously saw violet.

The reader may object loudly, maintaining that the red the observer now sees is identical to the red (R) previously located above the hole on the wall outside. His intuitive reaction would have been correct had the spectral colors been emitted directly from a light source (such as a lamp, beam AB), which then rotated toward the hole instead of being deviated by the prism. In fact, the direction of light from the actual source to the prism (AB) and the direction of the refracted light from the prism to the observer (Cr) did not change. The only changes occurred in the angles of obliquity of the refracting and dispersing interfaces of the prism and, hence, in whatever physical variables that correspond to these angles, mostly the intensity.

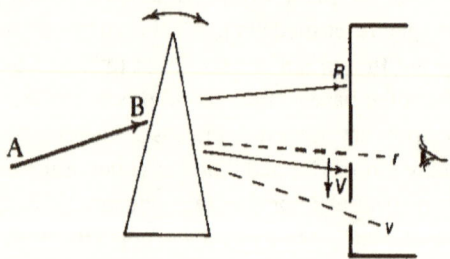

Meanwhile, the violet (V), which in the prism's first position was located in the upper hole, is now in the lower one (v); and it *looks the same* as when it was in the first hole. It may look the same and yet be physically different, as evidenced by the changed physical entities that generated it at the prism's interfaces.

The eye's poor discernment of intensity variations presented to it in slow temporal sequence is, by now, well-known and is related to its slow adaptability. To overcome it, the stimuli must be presented simultaneously side by side or in form of quickly alternating flickers. Only by these means (heterochromatic photometry) was it learned that the eye can actually discriminate between 7,500,000 different colors, whereas the number of words and theories that express our concepts of them is only in the dozens.[8]

The role of light's luminous intensity in changing the nature of the particular color perceived was recorded by Dr. Helmholtz, who noticed that with increased luminosity the perception of violet turned into a perception of blue or that of green into green-yellow.[9] Another clue is furnished by nature itself in the manner animals and man adapted to intensity variations. Nocturnal animals are provided with a sensitive retinal apparatus in order to function efficiently at night when the total quantity of light and variations thereof are small. These eyes are also most sensitive to the perception in the blue and violet part of the spectrum. Man's scotopic (night) vision is more sensitive to light that induces violet and blue than that which induces green.[10] The reverse is true under photopic conditions when the intensities of light are high—a phenomenon discovered and named after Jan Evangelista Purkyne (1787-1869). And as Newton wondered, "Was the Eye contrived without Skill in Opticks?"

Given that one is usually less impressed by evidence wrought from nature by simple experiments than by complicated and expensive schemes forwarded by men of some repute, the following observations by Edwin Land are of interest.[11] They originated with Maxwell in 1861 (when he combined red, green, and blue to produce all spectral colors) and were applied in 1894 to color photography by Louis and Auguste Lumière of Lyons.

He illuminated a multicolored scene, such as a group of fruits, with one color that corresponded to a very narrow strip of warm color from the red side of the spectrum, say yellow-orange, and then photographed the scene on a black-and-white film. The photographic transparency was formed by variations in the concentration of silver grains on a piece of celluloid; it possessed darker and lighter areas and could do no more than vary the *intensity* of transmitted light. He then photograph the scene again, this time illuminated with a cold color from the violet side of the spectrum, say yellow-green. The variations of black and white in this photograph differed ever so little from the first one because the yellow-green was somewhat differently absorbed and reflected by the scene than the yellow-orange.

He transmitted yellow-orange light through the black-and-white photographic transparency that was formed by intensity variations of this same light. He saw a yellow-orange scene. He then superimposed on this picture a projection in yellow-green light through a transparency formed by intensity variations of this light. The combined picture contained all spectral colors and more. He obtained all these colors by intensity variations of only two, two yellows in this case.

One must not, however, immediately infer from this that all these colors were generated by simple linear variations of intensity. When the intensity of two projected colors (the two yellows) is varied by using rheostats or density filters (but not photographs), no spectral colors ensue. This is not due to intensity randomness on the real scene as opposed to the order in the filter or due to retinal inhibition, retinal after-images, or other miraculous powers of the eye. The photograph receives the scene's light in subtle gradations, depending on the nature of the reflecting and absorbing subjects under illumination, and this subtlety cannot be duplicated by simple linear density filters. In the photograph, intensity variations of yellow light reflected off a red apple have no equivalent in a black-and-white density filter. The proper filter ought to be made by intensity gradations that result when one color-inducing light (say, yellow) is reflected off all the colors of the corresponding side of the spectrum.

To Edwin Land, who did the above experiments, the cause of these strange phenomena was in the eye: "The *independence of wavelength and color* [emphasis added] suggested that the eye is an amazingly versatile instrument." For he cleverly did not question the opinion of his predecessors concerning the nature of light itself: "This long line of great investigators cannot have been mistaken."[12] Perhaps.

The total quantity (intensity) of perceived light is made up of two components: First, since light manifests itself only in motion, intensity is given by the magnitude of this velocity, and secondly, by its brightness—the quantity per area. The velocity diminishes in transit from a rarer medium into a denser one, but this alone is insufficient to provoke color sensations, as evidenced by perpendicular transit. An oblique transit on the other hand diminishes in addition the brightness by enlarging the cross-section area, which, after refraction, is further enlarged by dispersion. The amount of light is also reduced by reflection. Above all, an oblique transit provides a gradient of these parameters, a slope, so that they differ in one side of the refracted beam from the other side.

In oblique transit from a denser medium into a rarer one, the velocity presumably increases. But in contrast to the opposite transit, the amount of light is already diminished by absorption before it reaches the interface. Furthermore, though the cross-section of the transmitted light is smaller, its enlargement by dispersion is much higher, as is the loss by internal reflection. Above all, the presence of matter renders the gradient in this transit steeper so that for beams of equal diameter the difference from, say, the left side of the beam to the right side is greater in the denser medium than in the rarer one; and upon transit, color sensations are that much sooner perceived.

When the function of the photoreceptive elements in the retina is reduced to that of simple transformers that convert photic stimuli into electric impulses in the neurons, the event is not dissimilar to the photoelectric effect. There the quantity of displaced electrons and their escape velocity, as manifested by the distance they travel (the total kinetic effect), is related to the total quantity of light absorbed. J. Elster and H. Geitel discovered in 1900 that the number of electrons emitted per second from a metallic surface was proportional to the intensity of incident light, and Lenard in 1902 turned this around and expressed intensity in terms of wavelength where the shorter waves elicited the greatest photoelectric effect.[13]

Kinetically speaking, the *violet* end of a spectrum is the slowest of color-inducing lights; its refractivity as well as specular and scatter reflectivity are the highest, which accounts for its *easy absorption*. In all, the violet-inducing light is the easiest to alter and transform. Hence, it most readily transfers its kinetic energy to the constituent parts of matter.

References

1. Fincham W. H. A.: Optics, 4th ed., London, Hatton Press, 1942, p. 243.
2. Brewster, D.: The Life of Sir Issac Newton. London, Gall & Inglis, 1890, p. 53.
3. Dollond, J.: Concerning an Improvement of refracting telescopes. *Phil, transact.* 1; 292, 1753.
4. Dollond J.: Of some Experiments concerning the Different Refrangibility of Light. *Phil. transact.* 50 (2); 733, 1758.
5. Wollaston, W. H.: A Method of examining refractive and dispersive Powers by prismatic Reflection. *Phil, transact.* 92; 365, 1802.
6. Reid, R. W.: The Spectroscope. New York, Signet, 1965, p. 67.
7. Huggins w. Further observation on the spectra. *Phil. Transact. Roy. Soc.* 158; 1868: 533.
8. Adler, F. H.: op. cit. Physiology of the Eye, p. 707.
9. Helmholtz, H. F.: op. cit. Ref. 21, p. 285.
10. Granit, R.: Sensory mechanisms of the retina. London, Oxford Univ., 1947, p. 123.
11. Land, E. H.: Color Vision and the natural image. *Proc. Nat. Ac. Sc.* 45; 115, and 636, 1959.
12. Land, E. H.: op. cit. Ref. 55.
13. Lenard, P.: Ueber die lichtelektrische Wirkung. *Ann. d. Phys.* 8;149, 1902.

Normal Spectrum

Allowing my concept and evidence of the prism's spectrum being composed of very many nanospectra, the question arises as to why is the topmost nanospectrum least refracted and the bottom one most refracted. An analytical approach to the answer will divide the prism into many horizontal sections (figure, left). The obliquity of the prism's refracting interface is the same in all sections; the only difference between them being the amount

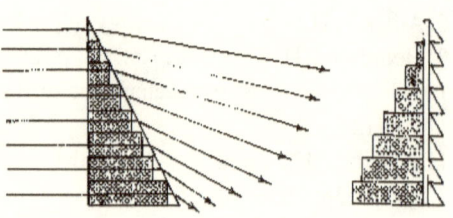

of horizontal matter through which light must pass on its way to the oblique interface. Traversing the longest distance through the medium near the base, the quantity of light lost by absorption and scatter is greater here than near the apex; its *intensity* on the active interface is lower here, and its refraction greater.

The distances separating the different colors in the final spectrum, the dispersion, evidently depend on the scattering and absorbing facility of these horizontal segments of matter and, accordingly, vary from one medium to another—irrational dispersion. It hinders any attempt at standardization because an event (such as a specific absorption line) that is seen through a prism made of one sort of glass in the green part of the spectrum may appear through another prism in the blue portion. To overcome this difficulty, a prism was constructed—along the lines of Fresnel's lenses—that was composed of many miniprisms in a straight line, thus eliminating the horizontal bars (right portion of right figure). The active dispersing interface remains constant without the disturbing variety due to the changing quantity of the medium (horizontal bars); there is hardly any refractive deviation here, except perhaps the minimal one caused by a thin plane-parallel plate. The width of the spectrum (dispersion) remains constant instead of widening.

The first to observe such spectra, produced by rills scratched unto a glass plate, was Claude Francois Milliet Deschales (1621-1678). The technique was then further developed by Joseph Fraunhofer and refined by Henry A. Rowland to create what is known as a *transmission grating*.[1] Dechales believed that since the colors were produced without a prism, they had nothing to do with refraction as Newton said but rather produced by variation in the intensity of light.

Light dispersed through this device is *not deviated* by refraction but forms a spectrum where the colors consistently occupy the same position, a normal spectrum, and where, therefore, an absorption line seen at the location of one color may be easily reproduced by any similar device irrespective of its material. Normal spectra may be formed by means other than refraction to deviate light from its course, such as diffraction or reflection, and we shall return to this later.

Grating devices helped standardize the spectrum and made its observation easier because there was no need to refocus the observation microscope on the different absorption lines. At the same time, they masked the spectrum's genesis; namely, its formation by deviation of light from its incident course as we outlined in the chapter on the parameters of refraction (neglecting to consider this origin tripped later observers to some curious cosmological deductions). In the case of transmission gratings, the spectrum is formed by refraction, but the concomitant dispersion is unaffected by the angle of incidence (ignoring deviation by plane-parallel plate). The effect is similar in Amici's direct vision prisms discussed earlier where the refractive deviation of one prism is annulled by the other.

In order to examine the spectrum in more detail, a greater separation between its colors is desirable; in other words, we wish to increase the resolving power of the device that produced the spectrum. To this end, we may either spread the dispersion by increasing the obliquity of the interface (thereby lowering the intensity per area) or else add incrementally to the horizontal bars on the way to the interface (figure) as explained above, which also reduces the intensity. As Lord Rayleigh put it, "That the resolving-power of a prismatic spectroscope of given dispersive material is proportional to the total thickness used, without regard to the number, angles, or setting of the prisms, is a most important, perhaps the most important, proposition in connection with this subject."[2] Increased resolving power by stepwise addition of material was first used by Michelson in the form of an echelon grating.[3]

REFERENCES

1. Rowland, H. A.: Preliminary Notice of the Results accomplished in the Manufacture and Theory of Gratings for Optical Purposes. *Phil. Mag.* 13; 469, 1882.
2. Rayleigh: Investigations in Optics, with special reference to the spectroscope. *Phil. Mag.* 8; 269, 1879.
3. Michelson. A. A.: The Echelon Spectroscope. *Astrophys. J.* 8; 37, 1898, and Proc. Am. Acad. Arts & Sc. 35; 109, 1899.

Anomalous Spectrum

By passage through a gradually increasing quantity of medium, such as given by a prism, the quantity of light reaching the refracting interface (AB in the figure) diminishes commensurate with this gradient; it is highest where after refraction the red-inducing end of the spectrum ensues, and lowest at the violet-inducing end. If we assign to the total quantity of incident white-inducing light the arbitrary numerical value of 100, for instance, its value on the oblique refracting interface may be $30 + 25 + 20 + 15 + 10$, made up approximately of the arbitrary products $(5.5 \times 5.5) + (5.4 \times 4.6) + (5.3 \times 3.8) + (5.2 \times 2.8) + (5.0 \times 2.0)$.

Suppose now that light prior to incidence on the prism enters a medium, say sodium vapor, that scatters and absorbs a very specific amount of its total quantity, say 20, whence it emerges diminished from 100 to 80 by the absence of, say, (5.3×3.8). When this light then passes through the graduated medium of the prism, it will reach the refracting and dispersing interface at one specific area greatly diminished. This area will be located where the diminution of light's quantity by the material of the prism matches exactly the prior diminution of the incident light by sodium, say (5.3×3.8). (The location of the area in the spectrum will of course vary according to the nature of the dispersing material, due to the irrationality of dispersion.)

Having thus been diminished, this light, at this location on the interface, will then be much more internally-reflective and refractive than the neighboring parts. Any *nanospectra* produced in this area will be more deviated (toward B', top figure) and will be missed

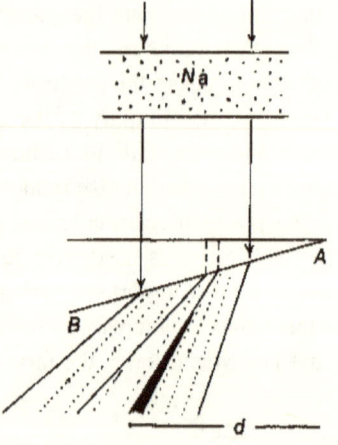

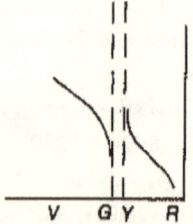

in the place where they would have ordinarily been in the normally deviated final spectrum, leaving what is known as an absorption line or absorption band (L), first seen by Wollaston[1] but examined in detail by Fraunhofer.[2]

By changing the angle of incidence, all the nanospectra become more, or less, deviated; and the distance (d) from the prism where the absorption line is seen changes accordingly. For instance, an increase in the angle of incidence increases the angle of deviation, the nanospectra overlap closer to the prism, and absorption lines will be noticed only when observing the spectrum in a plane nearer the prism. This forced Joseph Fraunhofer (1787-1826) to change the focus of his telescope as the deviation changed.[3] It furnished the impetus to replace the prism with Amici's direct vision prisms and grating devices.

According to Newton and Young's concepts, the absorption by sodium removed from incident white light a given ray of a given wavelength, say 590 nm; therefore, these waves were missing in the final spectrum. The difficulty was to explain how all these missing waves (or their neighbors) in different locations of the incident beam's cross section combined to one location in the yellow part of the final spectrum; in order to reach this one location, they must all be differently refracted, but if so, a coherent distinct line cannot possibly be visible right beyond the prisms. On the other hand, if one assumes that the waves occupy a specific location in the incident beam prior to dispersion, why then is the absorption line invisible without dispersion?

When a spectrum with an absorption line is again refracted, this time by a prism at a right angle to the first one, the ensuing spectrum does not form a continuous oblique image but is interrupted at the site of the line (bottom figure). As we approach the dark area from the less deviated red side, the spectrum suddenly becomes much more deviated due to diminished intensity by absence of the nanospectra, which deviated away. As we continue beyond the dark area, the spectrum is suddenly much less deviated due to addition of the more deviated nanospectra from the absorption area.

Instead of having light pass through absorbing matter prior to incidence on the dispersive prism, it may as well pass through it afterward as it would do when this matter (say, sodium) is incorporated in the medium of the prism or the prism made of it. When white-inducing light is cast on such a prism, a strange spectrum ensues. If the material scatters and absorbs yellow, for instance, it will cause the orange-inducing light (on the red side of yellow in the final spectrum) to be more refracted than the green on its violet side[4, 5, 6] (orange more refracted than green). This reversal of refractivity and color induction was termed anomalous dispersion where normal stood for Newton's and Young's views. (See experiment with the three separate spectra.)

REFERENCES

1. Wollaston W. H. A method of examining refractive and dispersive powers by prismatic reflection. *Phil. Trans.* 1802: pp 365-380.

2. Fraunhofer, J.: Bestimmung des Brechungs- und Farbenzer-streuungs-Vermoegens verschiedener Glasarten in Bezug aufdie Vervollkommnung achromatischer Fernroehre. *Ann. d. Phys.* 56;264, 1817.

3. Fraunhofer, J.: Kurzer Bericht von der Resultaten neuer Versuche ueber die Gesetze des Lichtes und die Theorie denselben. *Ann. d. Phys.* 74;337, 1823.

4. Leroux, M. F. P.: Dispersion anomale de la vapeur d'iode. *Comp. Rend.* 55;126, 1862.

5. Quincke, G.: Ueber die optischen Eigenschaften derMetalle. *Ann. d. Phys.* 119;368, 1863.

6. Christiansen, C: Ueber die Brechungsverhaeltnisse einer Weingeistigen Loesung Des Fuchsins. *Ann. d. Phys.* 141;479, 1870.

Redshift

Two distinct events are commonly referred to as a redshift, both in connection with effects produced by motion in the line of sight between a source and an observer. The first, the more puzzling and rare one, was reported by Herschel and others early in the nineteenth century concerning the color perceived from double stars. These binary stars apparently rotate in the plane of observation, and it was noted that the approaching star often had a blue tint while the receding one was tinted red. Christian Johann Doppler (figure) (1803-1853) in 1842 [1,2] formed a theory postulating that this variation of color was induced by the changed velocity of lights relative to the observer.

In Doppler's summary of his theory he said,

> If a luminous object, regardless of whether it radiates light or is illuminated by it, is moving directly towards or away from the human eye with speed related to the speed of light, this movement necessarily results in a change of the color *and intensity* [emphasis added] of the light . . . "When the speed of a moving star changes, its color and *intensity* also change, and so it may happen that in time a star appears to us in all the colors of the spectrum.

In 1848, Fizeau deduced from this report that the position of Fraunhofer's spectral absorption lines would be affected by the motions;[3] and Zöllner, with his reversion spectroscope, first proved in 1869 that indeed they were.[4] When the source approaches the prism—or any other spectrum producing device—the lines are visible closer to the violet end of the spectrum than when the source or observer are at rest. On recession, the lines are shifted toward the red end.[5] This second redshift phenomenon has recently received popular attention for, from it, the genesis of the entire universe was inferred.

Doppler's mode of reasoning as to the cause of the changed color of the moving stars must be understood within its milieu, for coming on the heels of the then common analogy of light to sound, particularly by Thomas Young, MD, it was also "strictly" analogical.[6] "What was said and claimed here for light waves also naturally applies very strictly to sound waves, and one therefore sought with good fortune all along to explain by way of analogy the various phenomena of light form those of sound." The pitch of a note is depressed when the emitting source recedes from the perceiving ear and is heightened when the source approaches. This was quickly verified by experiments conducted by Buys Ballot (1817-1890)[7] with a locomotive's steam whistle blowing along railroad tracks in the Netherlands and by Scott Russel in England.

If by comparison light is conceived as an undulating motion of material aether particles, as Young and Doppler did conceive, then recession would lower the velocity of propagation; and the waves would strike an observer at a lower rate—lower frequency—and, hence, appear longer. Furthermore, Doppler obviously accepted Newton's concept of white light being a loose mixture of all the colors of the spectrum separated by a prism like a sieve: "At such speed of the light-emitting object, if it were receding from us, the extreme violet and more so all other colored rays, consequently also the white light which is composed of them, no matter how intense, will become completely extinguished to observation." Somehow, recession slowed down only the red rays whereas advance accelerated only the violet ones. The longer waves were, at that time, assigned the power to induce the sensation of red. "It is known that perception of color is a direct result of pulsations or wave-impacts of the aether which occur at successive regular time intervals." In addition, the changed relative motion between the light source and its observer also affected its perceived intensity.

> In fact, nothing is easier to comprehend than that the distance and time interval between two successive waves must become *shorter* for an observer who is hurrying towards the oncoming waves and *longer* if he is moving away, and similarly, in the first case the *intensity* of the wave is stronger and in the second it must necessarily decrease. [Emphasis added.]

A few years after Doppler annunciated his theory, Hippolyte Louis Fizeau (1819-1896) (figure), for the first time, measured the speed of light when both the source and the observer were situated on earth (1849). His discovery a year later that the speed of light in denser media, such as water, was lower than in air fundamentally changed prior misconceptions about refraction. While proponents of the undulatory theory saw it as supporting their cause, for it contradicted Descartes and Newton's ideas about the mechanism of refraction which assumed a faster speed in the denser media, the fact that sound travels faster in denser media than in air actually militated against Young's and Doppler's analogies. Colored rays, or color-inducing rays, all

seem now to travel at different speeds in water, yet the sense of whiteness does not turn to red or any other color according to the refractive index of the medium. Doppler's analogy to sound did not account for the change of the star's total white-inducing light to only red or blue lights—rather than, say, green. On the same subject, Einstein, however, noted in 1916, "At all events we know with great exactness that this velocity [of light] is the same for all colours, because if this was not the case, the minimum of emission would not be observed simultaneously for different colours during the eclipse of a fixed star by its dark neighbour."[8]

Some have later attributed the phenomenon to variations in the star's temperature rather than motion.

When, late in the nineteenth century, a theory emerged that assumed a universal constant velocity for light (and therefore shrinkage of space and dilation of time), Doppler's explanation was turned around: light did not change velocity, but our time (referred to the velocity of light) dilated. Light, therefore, seemed slower to us. The frequency of its waves lower, their length longer, and therefore, we see red.

Experiment. A common phenomenon may illustrate the relation between luminous intensity of the source, its brightness, and the perception of its color. A wire, say in an incandescent lamp, is slowly heated by increasing an electric current flowing through it. At first, the wire appears red; and with increase in the current, its color changes to orange, then yellow, and finally white. We do not see the cold colors presumably because of superposition of the warm ones. In order to separate them further, we then repeat the experiment, viewing the wire through a prism. Again, at first we see red, and as the current increases, all other colors of the spectrum gradually appear in their turn, from orange to yellow, green, blue to violet. John William Draper found that when different bodies are heated, the spectrum is always generated in this order, and they all become white hot at the same time at the same temperature. The disappearance of colors with cooling is in reverse order.[9]

Wilhelm Wien (1864-1928) continued this line of investigation and arrived at what is known as Wien's displacement law.[10, 11] A black body, which, hypothetically, does not radiate or reflect any light on its own when heated, will emit light. The first light seen, at about 1700° K, is red. As the temperature rises, the other colors appear—the violet at around 2300° K. In addition, Wien noted that when all the colors are already present and the temperature rises further, it intensifies the violet end of the spectrum more than the red. And this helps explain the first Doppler effect.

65

We noted that in the spectrum, and in the normal dispersion curve, the deviation of the violet-inducing light increased disproportionately faster than that of the red as the angle of incidence grows, meaning that its velocity and intensity declined faster. Should the intensity or velocity of the total incident light be enhanced, violet induction will be proportionately more accented then red induction. When an emitting source of white-inducing light advances toward the observer, its *velocity* increases in reference to this observer, which increases its *intensity*; and the violet-inducing effect becomes disproportionately accented. The reverse occurs in recession—dim lights appear red.

The second Doppler effect was explained according to the undulatory theory along the same lines as the first: the frequency of light from a receding source declines, i.e., its waves becomes elongated—more red—and, therefore, the absorption line shifts to the longer-waved redder part of the spectrum.[12, 13]

Before proceeding to examine the second Doppler effect, a small matter of semantics deserves some clarification. When we observe a spectrum and say that we see in it a dark line, it is not the darkness that we see, but merely note the absence of light. Darkness cannot be seen—optically and visually, it is nothing. And nothing cannot do anything; in particular, it cannot shift. When the position of a given black absorption line within the regular order of the colors in the spectrum changes with motion of the source and is then noted in the spectrum's redder part, it is not the black line that changed but the whole spectrum. The spectrum has shifted to the violet side; that is, its light has been more deviated than that from a stationary source.

The first to discover and record, in 1802, the dark lines in the spectrum was William Hyde Wollaston, MD, (1766-1828),[14] a friend of Thomas Young, MD (figure). These absorption lines were then further investigated with a microscope instead of Wollaston's naked eye by the optician Joseph Fraunhofer (1787-1826). He also constructed a diffraction grating spectroscope, helping transform a qualitative method to a quantitative one while at the same time, unfortunately, masking thereby the origin of the spectrum in the deviation of the refracted lights.

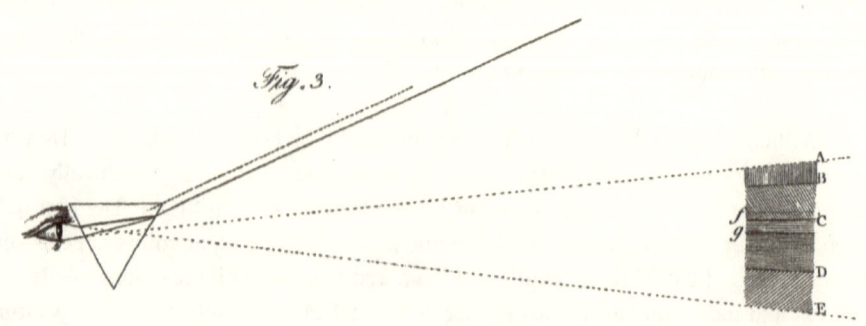

Fig. 3.

Applied to spectroscopic observations of moving celestial bodies, William Huggins said the following in 1868:

> If the stars were moving towards or from the earth, their motion, compounded with the earth's motion, would alter to an observer on the earth the refrangibility of the light emitted by them, and consequently the lines of terrestrial substances would no longer coincide in position in the spectrum with the dark lines produced by the absorption of the vapours of the same substances existing in the stars.[15]

Huggins's trials to confirm the Doppler effect resulted only in ambiguous data. At the same time, James Clerk Maxwell used the effect in a vain attempt to discover the value of the earth's daily motion by observing (on earth) if the refraction from a light source positioned on earth itself (instead of the stars) changed by its motion back and forth through two prisms, first in direction of the earth's daily motion from west to east and then opposite to the earth's motion (see "Michelson's Experiment").

As Kepler studied the eye in order to better understand astronomy, so may insight into the spectroscope ease our understanding of Doppler. The shift, the altered refractive deviation with motion of the source, cannot be noted when the observation is made through devices that are specifically meant to annul this deviation, such as Amici's direct-vision prisms or grating devices.[16] Huggins alluded to it in 1868: "If, instead of a spectroscope, an achromatic prism were used, which produces an equal deviation on rays of different periods, no difference between the light of different stars could be detected."[15] It will be further masked by limiting the field of observation to the immediate vicinity of the line as the reversions spectroscope did. These modern spectroscopes permitted more detailed analysis of the spectrum than was ever possible with a simple prism, yet as technological sophistication grew—and with it new knowledge—awareness of the primary cause of the whole dispersion phenomenon slowly receded into the background until it apparently completely disappeared.

The quantity of light falling on a given area of receiving interface is a function of three factors:

1. The quantity emitted by the source (its luminous intensity, its brightness)—the more emitted the more received.
2. The distance to the source. When the area of the source relative to its distance is small, the source is regarded as a point source—as would be the stars—and from such sources, the quantity received by a given area declines by about the square of the distance.
3. The relative velocity between source and observer. The quantity received per unit time is directly proportional to light's velocity; the slower the light, the less received per unit time.

We previously noted that light obliquely incident on an interface is the more refracted, the lower its velocity, or the lower its *intensity* per area interface, or both. Intensity is also lowered with distance from the source and its recession. When light's deviation—say, by refraction—has increased and absorption lines shifted to the spectrum's red end, the event is caused either by recession of the source (which lowers light's velocity) or by light's low density on the interface (which is a function of distance) or both. A redshift alone is insufficient to distinguish between them. For instance, we know since Bouguer that the intensity of light from the edges of the sun is lower than that from its center, and we also know that the redshift from the edge is greater.

Johann Karl Friedrich Zöllner (1834-1882) substantially advanced photometry beyond the works of Bouguer and Lambert with his newly invented (1865) very sensitive photometer he named *astrophotometer* (figure) that he used to measure light intensities from various sources, including the stars.[17] He found that the *intensity of lights from celestial bodies was proportional to their distance.* Furthermore, with his newly invented *reversionsspectroskop,*[18] he demonstrated for the first time that absorption lines indeed shifted with changed motion in the line of sight, as deduced by Fizeau from Doppler's theory. Zöllner's name is attached to a crater on the moon and to an optical illusion, but is seldom mentioned in the annals of science or optics. His obscurity is probably due to his strange borderline personality. In his book *Transcendental Physics,*[19] he ardently supported Henry Slate's spiritualism. He had an idea of a four-dimensional space, lashed forcefully against his perceived low moral standards of his colleagues, and was quite anti-Semitic.[20]

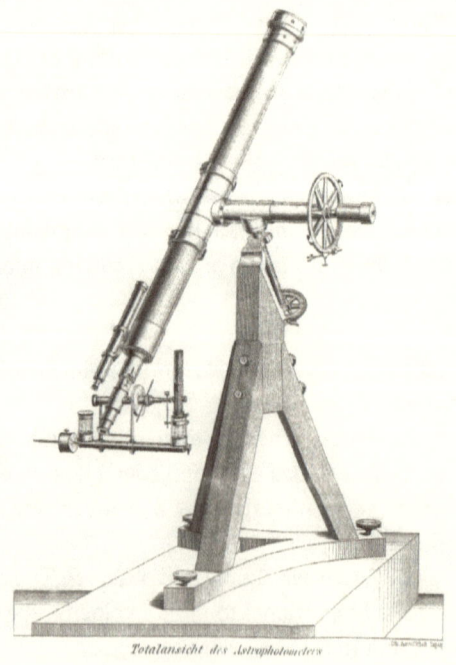

Totalansicht des Astrophotometers

Edwin Hubble noted in 1929 that the redshift of celestial bodies, seen from earth through modern spectroscopes, was also proportional to the distance of these bodies from earth; the farther the galaxy, the wider the shift.[21] Believing that a redshift implied recession only, the correlation meant that the farther the stars from earth, the faster they receded.[22] The universe as seen from earth was expanding. When one then traced the events back in time, the conclusion became inevitable that a compact single universe must have been born with one big explosion that propelled all its ingredients apart at an accelerating speed. One could pleasantly conclude from this that man on his earth was the center of the universe. "The unanimity with which the galaxies are running away looks almost as though they had a pointed aversion to us."[23] In addition, given the misery on earth, it looked as though objects that were already farthest away from it were nevertheless in the greatest hurry to escape farther.

This geocentrism, with its anthropocentrism, perhaps vindicated finally Aristotle and the church in Rome, but embarrassed less orthodox souls. Fortunately, it was quickly rationalized that position and motion have meaning only in reference to another position or motion, where the choice of naming the motion—the point of reference—was perfectly arbitrary.[24, 25] As Newton said, there was no way of ascertaining a favorable position; the earth was just as valid point of reference as the sun or any distant nebula. Even though the data were obtained from earth, it was not necessarily the center of the expansion.

The argument was perhaps irrefutable; the universe may be expanding even though no center is determinable, but this does not support the proposition, namely, that it truly is expanding. Similarly, the argument, for instance, that contraction of space and dilation of time with fast forward motion is undeterminable because our eyes, brains, and yardsticks all contract commensurately does not add weight to the notion that the event really occurs. Or the inability to prove that the world was not created at a given moment in biblical time (created with all evidence already in it, which supports the concept of its evolution over time) does not add validity or utility to the creationist idea that the world was really created at that time. Classical logic calls this form of faulty reasoning *argumentum ad ignorantiam*: ghosts must exist because nobody has established (or can establish) that they do not. [26] The earth is anywhere in the universe because nobody can prove that it is not.

Whatever secondary rationalizations followed the concept of an expanding universe, the primary optical data upon which it was originally founded seem inconclusive. Allowing photometric measurements and optokinetic understanding of the spectrum's genesis, it is hard to see how a redshift alone can distinguish a big bang from a small sigh. [27, 28]

References

1. Doppler, Ch.: Ueber das farbige Licht der Doppelsterne und einige anderer Gestirn des Himmels. Prag, Barrosch & Andre, 1842. Also in: Koenigl. Boehm. Gesel. d. Wissen. (5) abh. 2;465, 1841-42.

2. Doppler Ch. Beitraege zur Fixsternenkunde. 1846. Prague, [Gottllieb Haase & Sons]. Borrosch & André.

3. In a paper read before the Societe Philomatique de Paris, December 23, 1848, and first published in extenso in Ann. Ede Chim. et de Phys. 1870; xix:211.

4. Zöllner, F.: Ueber ein neues Spectroskop. Ann. d. Phys. 138;32-44, 1869.

5. Mach E. Beiträge zur Doppler'schen Theorie der Ton-und Farbenänderung durch Bewegung. Prag, J. G. Calve; 1873.

6. Doppler, Ch.: Bemerkungen zu meiner Theorie des farbigen Lichtes der Doppelsterne. Ann. d. Phys. 68; 1, 1847.

7. Ballot CHDB: Akustische Versuche auf der Niederlandischen Eisenbahn nebst gelegentlichen Bemerkungen zur Theorie des Hrn. Prof. Doppler. Pogg. Ann. 1845; 66: 321-351.

8. Einstein, A.: Zur Elektrodynamik bewegter Koerper. Ann. d. Phys. 17; 893, 1905.

9. Draper, J. W.: On a new Form of Spectrometer, and on the Distribution of Intensity of Light in the Spectrum. Phil. Mag. 8;75, 1879.

10. Wien, W.: Ueber den Begriff der Localisierung der Energie. Ann. d. Phys. 45;712, 1892.

11. Wien, W.: Temperatur und Entropie. Ann. d. Phys. 52; 132, 1894.

12. Gill, T. P.: The Doppler Effect. London, Academic Press, 1965.

13. Weinberg, S.: The first three minutes. New York. Bantam, 1979, p. 12.

14. Wollaston W. H. A method of examining refractive and dispersive powers by prismatic reflection. Phil. Trans. 1802; 92: 365-380.

15. Huggins, W.: Further observations of the Spectra of the Sun, etc. with an attempt to determine therefrom whether these Bodies are moving towards or from the Earth. Phil. transact. 158; 382, 1868.

16. Reid, R. W.: The Spectroscope. New York, Signet, 1965, p. 67.

17. Zöllner F. Photometrische Untersuchungen mit besonderer rücksicht auf die physische beschafenheit der himmelköreper. Leipzig. W. Engelmann, 1865.

18. Zöllner, F.: Ueber ein neues Spectroskop nebst Beitraegen zur Spektralanalyse der Gestirne. Pogg. Ann. d. Phys. 1869; 138: 32-44.

19. Zöllner F. Transcendental Physics. London; WH Harrison, 1880.

20. Zöllner F. Beiträge zur deutschen Judenfrage. Leipzig; O. Mutze, 1894.

21. Hubble, E.: A relation between distance and radial velocity among extra-galactic nebulae. Proc. U.S. Nat. Acad. Sc. 15; 168, 1929.

22. Hubble, E.: The Obervational Approach to Cosmology. Oxford, Clarendon, 1937.

23. Eddington, A.: The expanding Universe. Univ. of Michigan, 1962, p. 12.

24. Russell, B.: The ABC of Relativity. New York, Signet, 1958, p. 107.

25. Barrow, J. D., Silk, J.: The structure of the early universe. *Sc. Am.* 242 (4); 118, 1980.

26. St. Aubyn, G.: The art of argument. New York, Emerson Books, 1962, p. 104.

27. Silk, J.: The Big Bang. San Francisco, W. H. Freeman, 1980.

28. Singh S. Big Bang, The Origin of the Universe. New York, Fourth Estate; 2004.

Double Refraction

"There is brought from Iceland . . . a kind of Crystal or transparent stone, very remarkable for its figure and other qualities, but above all for its strange refractions . . . The first knowledge which the public has had about it is due to Mr. Erasmus Bartholinus, who has given a description of Iceland Crystal and of its chief phenomena."[1] The chief phenomenon described in 1669 by Dr. Bartholinus (1625-1692), brother of the famous physician Thomas, was that the crystal deviated incident light in two separate angles, resulting in two separate beams. He therefore termed in *doubly refracting* or, in modern usage, *birefringent*, and the ensuing beams *ordinary*, or usual, and *extraordinary*, or unusual.

Light conceived as refracted in two angles meant two refractive indices for the same medium; Iceland spar (calcite) was assigned the index 1.6584 for the ordinary ray and 1.4864 for the extraordinary one. It means that the two beams travel at different velocities and possess different intensities and different coefficients of reflectivity and transmissivity; when one beam is internally reflected and absorbed, the other may be transmitted. This last differentiation was accomplished in the combined prism of Nicol (1768-1851) by cutting the crystals in the appropriate angles.

The nature of light's asymmetry was conveniently resolved when Young suggested, and Arago and Fresnel agreed, that light must be constituted of transverse vibrations of aether particles, i.e., perpendicular to the direction of the flow instead of longitudinal. In a letter to Arago dated January 12, 1817, Young said,

> I have been reflecting on the possibility of giving an imperfect explanation of the affection of light which constitutes polarization, without departing from the genuine doctrine of undulations. It is a principle in this theory, that all undulations are simply propagated through homogeneous mediums in concentric spherical surfaces like undulations of sound, consisting simply in the direct and retrograde motions of the particles in the direction of the radius, with their concomitant condensation and rarefactions. And yet it is possible to explain in this theory a transverse vibration, propagated also in the direction of the radius, and with equal velocity, the motions of

the particles being in a certain constant direction with respect to that radius.[2]

Each such vibration could exist in a plane at any given angle in the beam's cross section. Ordinary light was a composite of vibrations in all angles, and Iceland spar separated a portion that vibrated in one angular plane from others that did it in planes otherwise inclined, similar as a prism separated composite white light into its constituent colors according to the length of the waves.

Since earlier in the nineteenth century, newer knowledge about the structure of matter disclosed asymmetry in the stereotectonics of some anisotropic crystals and soluble molecules—stereoisomerism. The idea that asymmetrical behavior of light may be due to asymmetry of matter rather than light was first suggested by John Herschel in 1822. This property of matter is given, for instance, to carbon atoms when they are variously bound to different radicals or, to a lesser degree, to the bipolar molecules of water. All these substances have in common the ability to more or less divide light into two components and are hence named optically active.

As we shall see in the next chapter, when periodicity produced by interference experiments was discovered, its cause was laid to the periodic nature of light—its longitudinal waves—rather than to the periodic nature of matter. Similarly, when polarization was discovered, its cause was seen in the polarity of light, this time its transversal waves. In the process, the nature of the aether had to be modified; for in order to accommodate the concept of swift transversal vibrations, as Young pointed out, the aether must not only be extremely elastic but also be a solid.

The internal structural configuration of anisotropic media may be altered by magnetic forces; and when this happens, it changes the light transmitted through them, for instance, the color perceived after transition through a doubly refracting crystal (Faraday effect)[3] or the color induced by a flame where the anisotropic matter acts as the source of light (Zeeman effect).[4] The electromagnetic forces while applied to matter were nonetheless often seen as acting directly on light. Had present knowledge of matter been available to Huygens, Newton, or Young, it would have probably answered to the cause of double refraction and polarization. As it was, it came too late.

On January 25, 1737, A.M. Rochon (1741-1817) demonstrated to the Paris Academy that when plane-parallel plates of different refractive indices were molten together, they became double refracting. Around 1809-1811, Malus, Biot, and David Brewster[5] discovered that polarization may ensue not only by double refraction but also by simple refraction, i.e., simply refracted light would traverse a doubly refracting crystal at only one angle of incidence; at other angles, it was scattered and absorbed. In their experiments, light was made incident at a great angle, between 80° and 90°, on a horizontal thin plate of glass where it was, as always, twice reflected and twice refracted in transition. After successive passages through between 8 and 47 such plates, the finally emerging light was found to be almost completely polarized, and the reflected and refracted beams behaved like the ordinary and extraordinary beams of a birefrigent crystal.

In order to better understand the phenomenon, we recall that when light obliquely passes a plane-parallel plate, it is reflected, *refracted at an angle*, and *dispersed* at the first interface, and then again more forcefully at the second interface. In the somewhat elongated cross section of the finally issued light (spectrum), one pole possesses different qualities than the other, and intensity distribution from pole to pole along the cross section varies accordingly. Repeated oblique passages through the plates diminish the total intensity of the transmitted and reflected beams while at the same time increases the asymmetry in their intensities, i.e., the spectral gradient steepens within each beam. Refractive and reflective dispersion (see there.) and the deviations by plane plates seem, therefore, to provide sufficient cause for light's asymmetry in the phenomenon—an asymmetry equal to the one in an ordinary prismatic spectrum. At a given angle of incidence to a doubly-refracting crystal, some lights (eliciting on occasion different color sensations) may be internally reflected and absorbed; at another angle, transmitted, just as the violet pole of one spectrum may be reflected, scattered, or absorbed while the red pole is transmitted.

We noted earlier that the angle of refraction, and with it the degree of dispersion, increases as the length of the path through the medium increases because of lowered intensity due to absorption. In a similar vein, Arago, in 1811, and Biot, in 1815, showed that the angle of polarization was a function of the thickness of the medium; and J. Seebeck, in 1818, showed that the angle depended on the concentration of birefringent matter when in solution.

A more cogent clue to the similarity of polarized light and the spectrally dispersed one is furnished by the perception of colors after polarization, as already noted in the Faraday and Zeeman effects. Giovanni Battista Beccaria (1716-1781), in 1762, discovered that most double-refracting crystals issued colored lights.[6] Were polarization merely a change in the angle of the plane in which the waves of light, or of aether, undulated, then it should not produce color sensations; for these were supposed to be due to the length of the waves. François Arago (1786-1813), in 1811, first reported the event where the spectrum's colors produced by double refraction through two successive birefringent crystals changed as one of the crystals was rotated through 180°. What was red in one position turned into violet in the second, and vice versa. He named the phenomenon chromatic polarization and pointed out the similarity to Newton's rings, where the transmitted and reflected colors were also complementary.

John Herschel later showed that red light was positively polarized ("turned to the right," or behaved like one of the rays issuing from a birefringent crystal) while violet was negatively polarized. And David Brewster (1781-1868) demonstrated how the colors of the spectrum—produced by reflective and refractive polarization through plane-parallel plates—changed to the complementary colors of the spectrum when the angle of incidence changed from either side of the angle of polarization. Later, we shall see that dispersion produced by reflection is complementary to that produced by refraction while more recent physiological data confirm the similarity, or identity, of polarization and spectral colors.[7]

In order to explain color sensations produced by polarization, Fresnel and Arago were obliged to employ the theory of interference, with it longitudinal waves, and suppose that polarized lights in crystals traveled at different velocities, whereby one wave overtook another and either destroyed it or changed its length and, hence, the color it induced. Which leads us to the next chapter.

REFERENCES

1. Huygens, Ch.: op. cit. p. 52.
2. Whittaker, E. T.: Theories of Aether and Electricity. London, Longmans, 1910, p. 121.
3. Faraday, M.: Experimental researches in electricity. London, Taylor, 1845, p. 2152.
4. Zeeman P.: On the Influence of Magnetism on the Nature of the Light emitted by a substance. *Phil. Mag.* 43; 226, 1897.
5. Brewster, D.: op. cit. p. 152.
6. Beccaria, G. B.: On the double refraction in crystals. *Phil, transact.* 52; 489, 1762.
7. Bernard, G. D., Wehner, R.: Functional similarities between polarization vision and color vision. *Vision Res.* 17; 1019, 1977.

Interference

On the twenty-fourth of November 1803, Thomas Young, MD, presented to the Royal Society in London his newly invented theory of interference, one of the more ingenious and helpful creations of the human imagination. The theory advanced two novel ideas: first, light beams traveling in space interact with each other, and secondly, this interaction may result in complete annihilation of all light. "Homogenous light, at certain equal distances in the direction of its motion, is possessed of opposite qualities, capable of neutralizing or destroying each other, and of extinguishing the light, where they happen to be united."[1] Sir George Airy later said, "We shall refer hereafter to experiments which show that the mixture of two pencils of light will produce darkness."[2]

Interference phenomena generally are not as common as those of reflection, refraction, or diffraction but are of high theoretical interest because, as Helmholtz put it, "Th. Young and Fresnel discovered the principle of interference, and primarily by dint of this discovery did the undulatory theory win general acceptance."[3] As opposed to Newton's corpuscular one. Or in Born and Wolf's words, "Historically, interference phenomena have been the means of establishing the wave nature of light."[4] And as Maxwell said, "Now we cannot suppose that two bodies when put together can annihilate each other; therefore light cannot be a substance."[5]

The term *interference*, in relation to light, was used by Newton simply to describe the disturbing effect of "intermingled," "overlapping," or "blended together" color-inducing lights, whereby the purity of color suffered and sometimes even turned to white. He formed no clear hypotheses regarding interactions between lights, leading to their enhancement or obliteration. Young, however, stipulated that as lights interfere in this manner with each other by superposition—by addition—they may also do so by subtraction. He followed Huygens in reasoning by analogy to sound and conceiving light as waves spreading from points on the luminous body in an elastic ethereal matter.

Ocean waves or sound waves, as manifestations of mechanical energy, were distinguished from the radiant energy of light; for whereas conversion of kinetic to potential mechanical energy was admitted, it was harder to see whereto the stored energy of light disappeared in the event of its destruction by interference without the invention of a material aether that was subject to the law's mechanics. The potential conflict was between three medical doctors. By the time Dr. Herman Helmholtz (1821-1894) and Dr. Robert Mayer (1814-1878) settled their priority dispute over the discovery of the law of conservation of energy,[6] Thomas Young, MD, was already gone; nonetheless, his theory survived this first law of thermodynamics for lack of a better one.

Classically, in order to produce interference phenomena, three conditions were usually prerequisite:

1. Light must come from a single source, preferably a small one.
2. The direction of part of this light must be altered, either by reflection, refraction, or diffraction.
3. The two parts must then meet in the same direction or at a certain angle of undetermined magnitude; for some puzzling reason, interference had not been obtained when beams traveled in opposite directions.

In practice, interference phenomena are often employed in the form of interferometers to measure distances. Since in this volume, thus far we have dealt mainly with deviation by refraction, we shall explore interference produced in this manner and add interference by reflection in the next chapter.

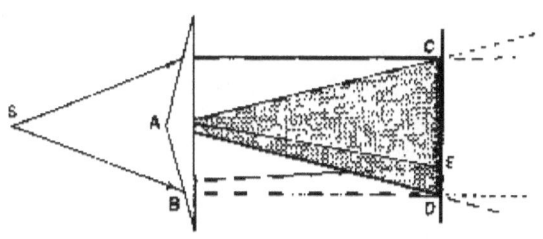

In Fresnel's biprism experiment, light from a small source is split in two by refraction through two prisms positioned base to base as illustrated. The ensuing spectra overlap (shaded area) and a screen placed at CD (figure) reveals bands or fringes of alternating bright and dim lights. According to Young, equally intense lights of the two spectra actually did travel in directions where almost no light was seen—darkness was positively produced by waves from one spectrum, annihilating the waves from the other spectrum.

Each of the two spectra is composed of many nanospectra, and the most deviated basal nanospectrum from the upright prism (ADE) (C - red; D - violet) superposes in reverse order of colors on the least deviated apical nanospectrum from the inverted prism (BDE) (D - red; E - violet) and so forth up the screen. The intensity value of each of the color-inducing lights in the nanospectra does not correspond one to another, and their combination results in intensity variations, manifested on the screen as bands of alternating bright and dim light. We recall that the dispersion of a spectrum is not linear; that is, the spread between the red end and the violet increases with distance because the violet is disproportionately more deviated. The spectrum and nanospectra produced by the upright refraction do not simply and congruently recombine with the

77

reversed spectrum to produce a white beam where light is evenly distributed, but as the various color-inducing rays of various nanospectra have different intensity values; their combination results in an image where the distribution of these intensities manifests periodicity.

The phenomenon of interference being in this case simply a redistribution of intensities by superposition of two incongruent spectra, it also ensues by successive transmission of a spectrum from an upright prism (A) through an inverted one (B) (figure). Though the two prisms may be of the same material and power, the second one—being at a distance behind the first that dispersed the beam into a spectrum—recombines the nanospectra in a different order, resulting in alternating bands of brighter and dimmer light. The top nanospectrum from the apex of prism A arrives at the base of prism B. In this experiment, one light suffered two deviations successively; there are no two lights interfering with one another, yet the periodicity effect (the fringes) is the same.

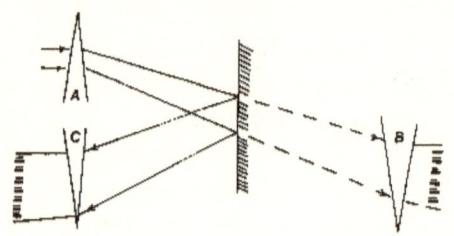

(Electron beams may also be deviated to produce an image that manifests periodic intensity distribution. And as Feyman pointed out, this image may also be obtained when a single ray of electrons passes *successively* through two slits down a screen; there is here no possible way electrons could interfere and annihilate each other.[7])

Let the first spectrum in the above experiment be reflected off a mirror and then refracted again through a second prism (C), which is the mirror image of the prism (B) in the previous experiment (figure). Again, we obtain fringes. When we substitute the base-to-base prisms with a convex lens, the circumstances are those of Newton's experiment that produced the interference circles called Newton's rings, first observed by Robert Hooke. We shall examine the reflective component of the phenomenon in the next chapter. The distribution of intensities in the normal spectrum is nonlinear; the violet end is disproportionately more deviated than the red, and this cannot be undone by the linear slopes of a prism placed upside down any distance from the generating prism. The recombined light, though often white inducing, is yet dilated; and intensity distributions in it are periodic.

Inasmuch as oblique transit through a plane-parallel plate also deviates light from its course, interference can be observed by means of such plates as we discussed when dealing with their refraction. Since dispersion in rarer-to-denser transit is less than that which ensues in the opposite transit, complete transit through the entire plate—first into it and then out again—does not cancel the dispersive effect, just as it does not cancel the refractive effect. And also like refraction, the dispersive effect would be the more obvious, the thicker the plate, and the greater the angle of incidence.

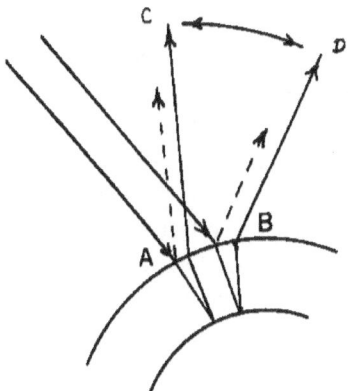

As seen in the illustration of a plane-parallel plate, part of the beam is reflected off the plate at a given angle of reflection. The other part passes through the plate twice (by internal reflection) and is deviated on exit; it emerges from the plate at a different angle than the reflected light. The two beams do not coincide in the same place but form alternating darker and lighter bands. Not because lights interfered with each other but because the nanospectra had different values and directions.

The effect is enhanced when a single small source is used to emit divergent rays and when the plate is curved convex to the impinging light (figure). The face of the plate not being uniformly inclined to the incident rays; the intensity at B is lower than at A. The path difference of the rays through the plate and their angles of emergence CD) also vary more in this curved plate than in a flat one, whereby periodic nanospectra become more readily visible.

Another arrangement was used in Michelson's interferometer: Light is made incident on a plate at 45°. One portion is reflected from the plate, then returned by a mirror (p. 68 and 112) and again reflected in direction of its origin. The other portion is refracted by the plate and then returned by a mirror through the plate. Since the transmitted light is deviated by the plate, it returns at an angle to the reflected untransmitted light, and the two do not coincide in place but form brighter and lighter interference fringes.

The width and distance between these bands may be varied by moving the mirrors back and forth from the plate, thereby altering the distance between the two transitions through the plate. The fact that the width of the fringes and the distance between them varies with distance of the screen from the generating interfaces proves that the rays exit at an angle to one another. When one expected the reflected and transmitted lights to return to the same location, one needed Young's hypothesis of interference to explain away the light missing between the bands.

In 1927, Clinton Joseph Davisson (1881-1958) and Lester Germer of the Bell Laboratory in New York discovered that electrons that passed near or through matter (reflected or diffracted) exhibited effects similar to those of light, i.e., bands of higher and lower intensity. Prince Louis-Victor Pierre Raymond de Broglie (1892-1960) inferred that electrons must propagate in an undulatory fashion and (like Young's light) self-destruct on the way.[8, 9] First, Maxwell said that light cannot be a substance because this could not possibly self-destruct, so light was a wave of energy; but then Broglie said that the substance of electrons was also a wave (wave-mechanics) and does self-destruct without leaving any effects.

Interference phenomena produced by refraction are explicable when account is taken of angular refraction by plane-parallel plates, the dispersion produced by reflection, and the nature of the spectrum as composed of nanospectra, obviating the need to imagine the self-destruction of lights in space. The periodicity attributed to the nature of light itself in order to account for the phenomena appears to be due to nonlinear intensity variations in the spectrum produced by the action upon light of *periodically distributed matter*; that is, its composition of discrete molecules and atoms.

REFERENCES

1. Young, T.: A course of Lectures, op. cit. pp. 613-630.
2. Airy, G. B.: Undulatory Theory, op. cit., p. 13.
3. Helmholtz, H. F.: Handbuch, op. cit., p. 308.
4. Born, M.: op. cit. p. 256.
5. Maxwell, J. C.: Ether. In: Encycl. Britt. 9th ed., vol. 8, Chicago, R. S. Peale, 1892, p. 568.
6. Helmholtz, H. F.: Vortraege und Reden, 4th ed., Braunschweig, Vieweg, 1896, vol.1: 401
7. Feynman, R. P.: The Character of Physical Law. MIT Press, Cambridge, Mass., 1967.
8. Broglie, L. V.: A Tentative Theory of Light Quanta. *Phil. Mag.* 47; 446, 1924. Also: *Ann. d. Phys.* 3; 22, 1925.
9. Broglie, L. V.: La mechanique ondulatoire des systemes decorpuscules. Coll. de Phys. Math., Fasc. 5, Paris, 1939.

Reflection

1. REFLECTIVITY

We return to the question of why do some lights break through an interface between transparent media and are refracted while others are reflected back? We noted earlier that reflectivity and refractivity are positively correlated; in Born's words, "It can easily be verified that, in agreement with the law of conservation of energy, $R + T$ =1." (Where R is reflectivity, and T is transmissivity.) Refractivity is also correlated to the ratio of velocities at the interface—the higher this ratio, the greater the refraction. Correspondingly, reflectivity also varies with the ratio of velocities: specular reflection off an interface does not occur without appreciable difference in the velocity of light in the two media. The higher the velocity differential on the interface, the higher the reflectivity, and the lower the transmissivity.

The total *quantity* of the transmitted portion and, hence, its intensity is the lower the higher the reflectivity, i.e., the higher the quantity reflected off the interface. This reflectivity increases with the angle of incidence. At the same time, the *intensity* of the transmitted light is also the lower the slower its *velocity* in the medium. The low intensity of the transmitted light produced by these two factors leads to its high refractivity and reflectivity, as noted when discussing the reflective component of refraction.

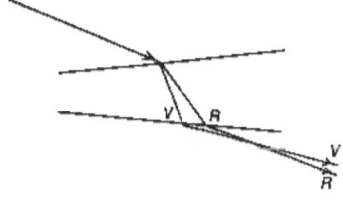

Experiment. The events are evident in the following arrangement: I cast a narrow beam of light upon a small angled prism, say 10°, from direction of its apex and observed the spectrum on a distant screen. The prism was then slowly rotated clockwise (figure) until the violet disappears from the screen; with further rotation, the other colors disappeared in turn, blue to green to red.

The violet end of the spectrum (V) formed by the prism's first interface was, as always, more refracted and, hence, more perpendicular to the prism's second interface than the spectrum's other end perceived as red (R). With rotation of the prism, the violet eventually disappeared from the screen by dint of total internal reflection. This occurred despite the violet's more perpendicular incidence on the reflecting second interface than the red or, in other words, under equal angles of incidence the violet-inducing pole of the spectrum is more reflective than the red and the red more transmissive.

Evidently, under *equal* angles of incidence different parts of the spectrum are differently reflective; therefore, reflectivity, particularly internal, cannot be a function of the geometrical angle but of some other physical feature peculiar to the various lights. As Newton put it, "Those Rays are more reflexible than the others which are more refrangible." And as the velocity of the more reflexible lights in the denser medium is lower than that of the less reflexible ones, one may conclude the following: at an interface, the lower light's velocity, the higher its reflectivity; and the higher the velocity, the higher the transmissivity. That is, transmissivity is directly proportional to light's velocity whereas reflectivity is inversely proportional to it.

The intensity of a light beam is composed of two components: (1) the velocity of light—quantity in time, and (2) its density per area—quantity in space. The higher velocity of the red pole, which accounts for its transmissivity at a certain given density on the interface, may be counteracted by enlarging the area. This is accomplished by increasing the angle of incidence, which eventually leads to the red's total reflection as the violet did at lower velocity but higher density.

The following two experiments by Newton exhibit the circumstances. In his experiment 9 (figure), light F is perpendicularly incident on wall AC of a right-angled prism. The spectrum HG is observed. The prism is then slowly rotated clockwise in direction ABC and the reflected light N refracted by a second prism. "I observed that when those Rays, which in this [ABC] Prism had suffered the greatest Refraction, and appeared of a blue and violet Colour began to be reflected, the blue and violet Light on the Paper, which was most refracted in the second Prism, received a sensible Increase above that of the red and yellow, which was least

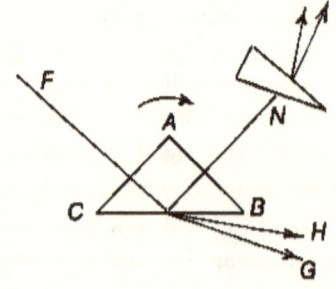

refracted; and afterwards, when the rest of the Light which was green, yellow, and red, began to be totally reflected in the first Prism, the Light of those Colours on the Paper received as great an Increase as the violet and the blue had done before." (Unless Newton used water prisms—which, in this case, is doubtful—the actual experiment cannot be reproduced as told because glass will have already totally reflected all light from the prism's base before the experiment had begun at 45° incidence; in modern terminology, he fudged the experiment. However, a right-angled prism is actually not prerequisite to demonstrating the truth of Newton's final conclusions.)

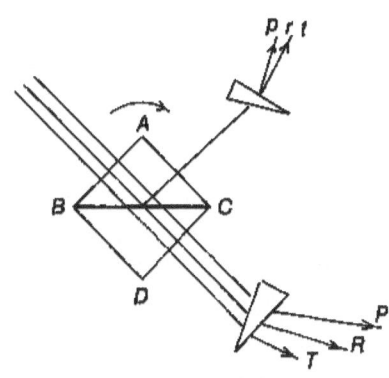

The transmitted light may be analyzed according to Newton's next experiment (No. 10). Light is transmitted through two right-angled prisms tied base to base and is then passed through a third prism (figure). When the two prisms ABCD are rotated, the violet is the first to disappear from the transmitted spectrum PRT while being enhanced on the reflected spectrum prt.

The two prisms in Newton's experiment form a parallelepiped with discontinuity BC where their bases touch. By introducing further such discontinuities in the manner illustrated in the figure, light is selectively reflected and transmitted by each one, whereby the reflected portion eventually becomes totally trapped in the medium. The finally transmitted light lacks that part of the original incident light which was, in this manner, absorbed. When light

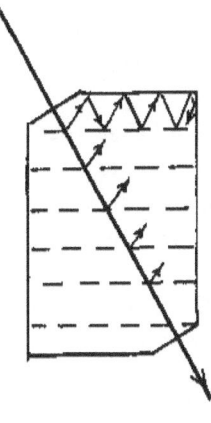

passes through a body where it is repeatedly internally reflected, by specular or scatter reflection, and whence it emerges diminished by that portion, the event may be termed *absorption*. One may safely conclude from the above that the violet is more readily absorbed—is more absorbable or has higher *absorptivity*—than the red, a phenomenon commonly seen on foggy days when red lights are visible to greater distances than blue ones or when violet more readily causes superficial skin and corneal burns whereas the red causes deeper ones. In opaque bodies or metals, absorption is not as easy to inspect as in transparent media, but neither substance can reflect light without some loss of light by transmission and absorption.

Methods meant to measure the quantity of light according to its effect (photometry) must take into account the various absorptivities of lights; a more absorbable light may lead to a higher reading on some instruments (measuring, for instance, the effect on the skin's surface or the effect on electrons in the surface of metals) even though optokinetically speaking it possesses a lower value of kinetic energy.

In this connection, it is interesting to note that retinal rods and cones are structured in multiple layers. When light traverses these plates perpendicularly, and thus presumably passes through more numerous layers, light produces an appreciably more intense sensation than when it traverses them at an angle (Stiles-Crawford effect).[1] In addition, retinal elements for the discrimination of colors (the cones) are located in the posterior hemisphere of the globe where light impinges more perpendicularly, whereas the rods—meant for detection of intensity variations only—are located peripherally.

Newton explained variable reflectivity and transmissivity of different lights by conceiving them as "light globules" periodically pushed through the interface:

> Nothing more is requisite for putting Rays of Light into Fits of easy Reflexion and easy Transmission, than that they be small Bodies which by their attractive Powers, or some other force, stir up Vibrations in what they act upon, which Vibrations being swifter than the Rays, overtake them successively, and agitate them so as by turns to increase and decrease their velocities, and thereby put them into these Fits.[2]

The impact of light corpuscles stirs the medium into vibrations that, according as they are in phase with the bodies or opposed to their motion, either push them through the interface or reflect them back.

The undulatory hypothesis of Huygens and Young envisioned light as a *longitudinal* vibration, not of the medium itself but of the ethereal matter within it. "That a medium resembling, in many properties, that which has been denounciated ether does really exist, is undeniably proved by the phenomena of electricity; and the arguments against the existence of such an ether, throughout the universe, have been pretty sufficiently answered by Euler."[3] When aether particles reached the interface in an energetic compressed state the wave was transmitted, whereas in the weak expanded state it was reflected.

After Young suggested that the vibrations of the aether particles be viewed as *transversal*, i.e., they were undulating sideways in a plane perpendicular to the direction of light, similar to water waves—not those inside the water, but those on its surface. It helped Fresnel and Arago rationalize their observations and thoughts on the nature of polarization. When such a sinuous transversal wave reached the interface in its perpendicular phase, it was transmitted; when it was incident at some other (undetermined) angle, it was reflected (figure).

Finally, when the existence of a luminiferous aether was cast in doubt by Michelson's experiment, the need for a new theory arose. In its absence, one resignedly accepted the undulatory theory as a convenient mathematical model satisfactorily representing to a susceptible intellect the known facts of optical reality. A simpler (optokinetic) interpretation to variable reflectivity and transmissivity suggested itself to me by data which established that light exerts pressure, i.e., it has momentum, and that different lights travel at different velocities so that a faster light breaks through an interface while a slower one is reflected.

2. ANGLE OF REFLECTION

The angle of incidence and the angle of reflection are both traditionally measured from the normal to the interface (α in figure). It is, however, somewhat simpler and more to the point to focus on the angle of *inclination* ($\beta = 90 - \alpha$)—as we noted also in relation to refraction—because incidence is to the interface itself and reflection is off it rather than the geometrical normal. Additionally, since the act of reflection involves a change in the direction of light, it is useful to talk about an angle of reflective *deviation* (γ)—the angle that a reflected beam forms with the direction which the incident beam would have had beyond the interface.

> Quoniam autem et refractiones faciant in angulis equali-bus in speculis planis et circularibus, per eadem demonstra-bimus, celeritate enim incidentie et refractionis. necessarium est enim rursum per ipsas minimas rectas conari. dico igitur, quod omnium incidentium et refractorum in idem radiorum minimi sunt, qui secundum equales angulos in speculis planis et circularibus si autem hoc, rationabiliter in angulis equalibus refringuntur.[4]

This quotation is from Hero of Alexandria (c. AD 62) and conveys the idea that since the act of reflection is very fast, the rays necessarily do it in the shortest way, and the shortest way in reflection is given when the angles of incidence and reflection are equal.

That this fundamental principle has not been changed for almost two thousand years is perhaps one of the more melancholy states in optics.

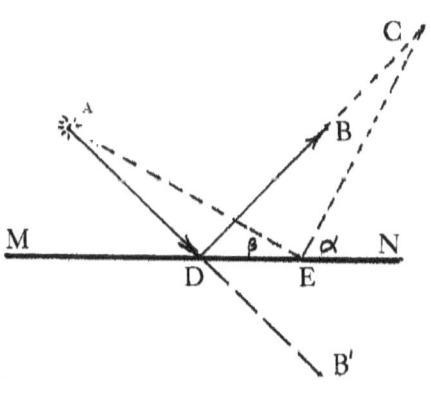

Newton wrote, "Axiom II. The Angle of Reflexion is equal to the Angle of Incidence." And Mach,[5] for one, phrased it more precisely as illustrated: "Light on being reflected at the plane mirror NM moves from the initial point A to the final point B via point D in such a way that the path ADB is shorter than any other path which passes through any other point of the plane." The shortest path is a straight line AB', and since B' is the mirror image of B and the path ADB' equals ADB, it follows that the angle of reflection equals that of incidence. This all tacitly assumes that the velocity of incident light is the same as that of the reflected one, or is instantaneous. So much for Euclidean geometry.

85

As we noted also in connection with Fermat on refraction, the difficulty of applying teleological reasoning to reflection is that the only known fact is the angle of incidence on the mirror; light does not know that it wants to go to final point B and that it wants to do it in the shortest time. That light goes to B, or anywhere else, is the proposition to be proven. Hero proved it from the general premise that anything imbued with a strong determination to travel "fast" would do it at a minimum length of path. The applicability of such general principles to particular events is not always self-evident. (See also "Motion" under "Definitions.")

Of all the endless possibilities, Hero's axiom applies solely to the one particular case where light's source (A) and its reception (B) are equidistant from the plane MN. Suppose, instead, that the light goes to position C. DC is longer than DB and takes more time to transverse. In order to minimize the time traveled from A to C through D, this path needs to be shorter; that is, it must reflect on the plane at position E where EC is closer to the vertical (the shortest distance from a point to a plane), and here, the angle of incident does not equal that of reflection ($\alpha > \beta$). If we assume for a moment that light after reflection at point D is retarded (speed AD > DB), the path of least time will also need to be through position E away from D until ultimately angle $\alpha = 90°$ which, as we shall see, is probably contrary to actual events.

According to Galileo's and Newton's principle of inertia, anything moving tends to do so in a straight line, and any deviation must therefore be attended by external forces that will effect a change in velocity. When no force (no energy) is added, a deviation from the original straight path is inevitably coupled with diminished velocity; nothing, material or not, may thus deviate from a linear course and still maintain its original velocity. Indeed, it would be astounding if deviation by means of reflection did not involve a change in velocity (as cause or effect) or if a change in velocity did not alter the magnitude of the reflective deviation.

At incident inclination near 0° (a right angle of incidence), light grazes the interface, and the reflective deviation is nearly zero. At the other end, at perpendicular incidence, the angle of reflective deviation is 180°—light completely reverses its direction. We know for a fact that the angle of deviation by reflection diminishes as the angle of incident inclination diminishes; for a given width of beam, a smaller angle of inclination means a growing cross-section area at the interface and, therefore, diminished intensity per area of interface. At the same time, for a constant angle of incidence, the intensity on the interface can be reduced by two other circumstances; first, by a slower velocity of incident light referred to the interface (or slower velocity of the interface referred to the light) and, secondly, by a lower velocities ratio on the interface, i.e., the velocity behind the interface (and after refraction) approaches the velocity prior to reaching the interface (i.e., the media have similar indices of refraction). We thus have three causes that may be expected to result in a lower angle of reflective deviation: (1) A growing angle of incidence. (2) Lower velocity of incidence. (3) Lower velocities ratio. All three mean lower intensity on the interface.

In a situation similar to the one that led Bradley to discover aberration or to Michelson's experiment, take three observers, one (C) is on a moveable platform while two are stationed outside the platform at A and B some great distance away (figure). They have decided upon a moment in time at which to start the platform moving (v) and when on this platform shall light be emitted to be reflected from mirror M on the platform. The question is who will receive the reflected light, A or B?

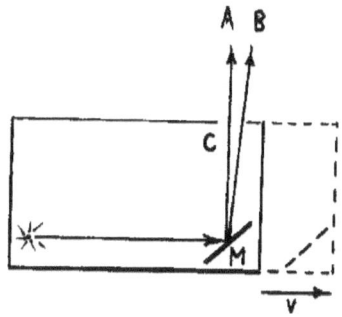

Or let an observer (C) with a light source be inside a car moving at velocity v. He notes the reflection off a mirror inside. Light travels at velocity c referred to the observer and is reflected to point A, say perpendicular to the direction of incidence. However, to an outside observer in reference to whom the car moves, point A moved to B by the time light from M reached it. To him, therefore, light deviated from its projected course at an angle smaller than a right one (larger angle of reflection). To him, the original light moved forward at velocity v + c. But since the mirror moved with the car in direction of the light, the observer records light's velocity at the receding mirror as c, i.e., slower than v + c. Thus, when light's velocity relative to a mirror is diminished (c versus v + c), the angle of reflection grows (its deviation from its incident direction is lessened) and vice versa.

Actual measurements of angles of reflection off a very speedily rotating mirror revealed that when rotating in direction away from the light, the angle of deviation declined, and vice versa; when the mirror was on approach to the light, the deviation increased. The effect of a changed velocity of incidence on the angle of reflection was first investigated experimentally with moving reflecting prisms and with moving mirrors by Sagnac [6] and others;[7] it was more recently confirmed by Kantor.[8, 9] The data are in accord with the optokinetic concept that for a given angle of incidence, the angle of reflection is a function of the relative velocity of light to the mirror. When a mirror moves in the same direction as that of incident light, thereby decreasing light's relative velocity, the angle of reflection increases (deviation declines); it decreases when the mirror approaches the light source, thereby increasing the relative velocity of impact. Arnold Sommerfeld pointed out a fitting mechanical analogue: "A tennis ball which falls obliquely on the racquet is reflected at a smaller angle than that at which it impinges. This is because the perpendicular component of the ball's velocity is increased by the forward motion of the racquet."[10] The analogy is somewhat limited by the fact that light striking an interface is also transmitted, which should not occur to the ball striking a good racquet.

The slower the speed of light at the reflecting interface, the less its deviation by reflection; that is, the nearer to the reflecting body the reflected ray lies. At the same time, the slower the light, the higher the reflectivity, and the greater the deviation by refraction, which here too approaches the center of the refracting body. As we shall see under diffraction, light also deviates toward material bodies that are not transparent—a deviation that used to be termed *inflection*. It hence looks as if material bodies, transparent or not, attract light; and the slower the speed of light, the greater this attractive effect.

The change in velocity of light in reference to a moving interface or mirror, or the change in the velocity of the interface in reference to the light (which since Oresme and Copernicus is accepted to be the same), brings us in conflict with a widely held notion that the velocity of light in moving frames of reference is a universal constant, and therefore, we deal with this idea at length in the optokinematic part.

A receding mirror means that the intensity per minute of light impinging on it is lowered; a higher angle of reflection than the angle of incidence means a lower angle of reflective deviation and a smaller cross-section area of the reflected beam compared to a beam reflected at the angle of incidence, which, in turn, means higher intensity per cross-section area in this beam. In the figure angle $\beta > \alpha$, and $a > b'$, while b and b' are *nearly* the same. The smaller angle β, the smaller is area b'. The less light is deviated from its source by reflection, the higher the intensity per area in the reflected beam, which is also empirically true when the mirror is stationary and the angle of incidence grows, as first shown by Bouguer.

Similar circumstances apply in the event of refraction. There, for a given angle of incidence, the deviation is the smaller the less light's velocity of incidence is altered (low refractive index), which means low intensity on the interface. The angle of deviation by reflection or refraction apparently depends on the intensities ratio on the interface; the lower this ratio, the less the deviation, and vice versa.

The space in which light moves is rendered asymmetrical (anisotropic) by an interface—one part is occupied by a denser medium than the other part. This asymmetry is the lesser, the lower the velocities ratio at the interface (similar refractive indices). For a given angle of incidence on an interface, the deviation by refraction will decline as the velocities ratio declines. Since a

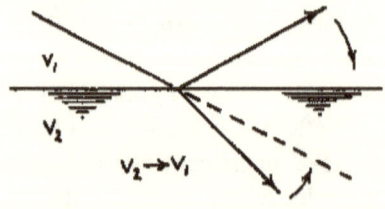

lower ratio means lower intensity on the interface, the angle of deviation by reflection also declines. That is, the more asymmetrical the space, the greater the angle in which incident light splits, and the more isotropic, the less the separation.

3. REFLECTIVE DISPERSION

When obliquely incident white light is deviated from its original course by refraction, the refracted rays diverge one from another—dispersion—and often provoke color sensations, though not always readily, as Goethe observed. Obliquity of incidence leads to different lengths of path on each side of the beam (left and right in the illustrations), different times of arrival at the interface, and different intensities on the interface. It results in different intensities along the opposite sides of the refracted and reflected beams as viewed in cross section on the plane of the interface. We spoke of refractive dispersion and will expect reflective dispersion as well.

The first to record reflective dispersion off metals and glass was Deschales in 1674. When light is incident on glass, the reflected portion complements the refracted one, for the two comprise the total $(R + T = 1)$; $(R = 1 - T)$ where T stands for transmissivity and R for reflectivity, "in agreement with the law of conservation of energy." If then the refracted beam is graduated from pole to pole (the spectrum), the reflected beam ought to exhibit similar features in complementary order. We already pointed out several instances where the reflected and refracted lights acted in a complementary manner. For instance, with a growing angle of incidence, the angle of deviation by reflection diminishes whereas that of refraction grows; and the intensity of the reflected beam grows whereas that of the refracted one in the medium diminishes. In the reflected or in the refracted spectrum (created perhaps by corresponding gratings), absorption lines of a receding light source will shift to the red end of the more deviated spectrum.

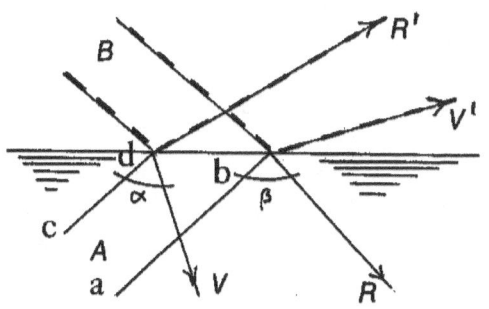

Consider first the case in the denser medium (figure). Beam A arriving at the interface from the denser medium is dispersed into spectrum VR, the violet end V being more reflexible than R (beam B from the rarer medium is reflected, refracted, and dispersed into VR and V'R'). On transmission, beam A forms the spectrum V'R'. The reflected and refracted spectra are complementary, and furthermore, the order of colors in the internally and externally reflected spectra is reversed; the externally reflected spectrum corresponds to the refracted spectrum produced by light that arrives at the interface from the inside, and vice versa. The angle of deviation by reflection declines as the reflecting interface or the light source recede from one another because light's velocity declines. The ensuing reflective spectrum R'V' shifts toward its violet end, and any absorption lines seen in a stationary position will shift to the red end.

The letters V and R in the illustration mean violet- and red-inducing lights. The actual occurrence of these sensations depends on the experimental circumstances. For color sensation is a function not only of the intensity, as DesChales believed, but also of the intensity gradient in the spectrum from one pole to the other; that is, the angle of dispersion. As is the case with refractive dispersion, the reflective dispersion is greater when light arrives at the interface from the denser medium where its velocity is lower, whence the gradient steeper.

White light refracted and dispersed through a weak (small-angled) prism still appears white very near behind the prism. The spectrum is visible only at some distance beyond the prism or only when the incident beam is very narrow. Similarly, reflective dispersion is more readily visible when the reflected beam is narrow. This occurs when a fine groove is scratched into the surface of reflective glass or metal, thus creating a very narrow surface that is at an angle to the rest. An enhanced and more normal reflective spectrum is obtained when multiple grooves are scratched into a surface, as is employed in *reflection gratings*.[10] Dispersion is also enhanced when the reflective surface is convex to the incident light so that the reflected rays diverge, as is the case in refractive dispersion through plane parallel plates. Such reflective dispersion is visible when light is incident on a fine hair or parts of a feather. The phenomenon manifests itself also when light strikes convex surfaces of highly transmissive thin films, such as *soap bubbles* (see figure). The orthodox interpretation of these phenomena required Young's assumption of interference.

Newton, who considered only refractive dispersion, interpreted the rainbow in these terms. The primary bow was created by refractive dispersion in and out of the rain drops with one internal reflection; the secondary bow was similarly created—by two refractions and triple internal reflections (of equal angles). With due respect, nonetheless, available theoretical or empirical evidence is not overwhelmingly against the view that the *rainbows* are simply created by reflective dispersion though the matter is but of cursory interest considering the length of time we spend outdoors marveling at rainbows.

Similar as the total refractive spectrum produced by a prism is the sum of nanospectra, so also the reflective spectrum. But the nanospectra produced by reflection do not correspond to those produced by refraction and are more likely to be complementary. It is thus possible to recombine, to some extent, the dispersed refractive nanospectra by simple reflection.

Experiment. I let a spectrum produced by refraction through a prism be incident at a high angle onto a glass plate. Were the angles of the spectrum's rays not variously changed by reflection, the spectrum ought to continue dilating uniformly and appear after reflection just as it did without it, a mirror image. In fact, however, on a screen near the plate, the spectrum appeared as discontinuous bright and dim bands (figure). These bands were not produced by any imperfections in the refracting or

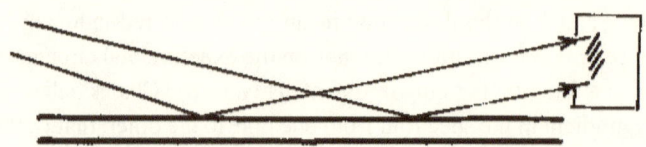

reflecting media, for when I tilted either of them in one plane or another—thereby tilting the spectrum—the bands did not tilt correspondingly.

The result is similar to that created by Lloyd's mirror where the primary spectrum was produced by diffraction.[11, 12] Lloyd's results were explained by assuming that the reflected light interferred with direct light flowing from the slit to the screen. But in my demonstration—as previously observed in regard to Fresnel's biprism—no direct light fell on the screen, i.e., the phenomenon required no other light to produce it and no hypothesis of interference to explain it.

One late afternoon in 1808, Étienne-Louis Malus (1775-1812) was looking from his room through a doubly refracting crystal at sunlight reflected off a windowpane of the Luxembourg Palace in Paris and discovered that by rotating the crystal, he extinguished one of the double images of the reflected light.[13] He thence concluded that reflected light was also polarized because it had the nature of either one of the beams issued by a doubly refracting crystal. Similar as the ordinary or extraordinary ray of double refraction could be extinguished by the proper angular incidence on another doubly refracting crystal, so too could the reflected ray be extinguished by another reflecting plate positioned perpendicularly to the plane of the first reflecting plate. Reflected light has since been regarded according to the undulatory hypothesis as polarized, though it may also be understood without this assumption as follows.

A beam obliquely incident on a transparent medium, such as window glass, loses on impact part of its intensity to transmission (or absorption), a loss which varies from point to point along the elongated plane of incidence; intensity diminishes also by reflective dispersion, which also varies across the plane. When this reflected beam then strikes an interface at right angles to the first interface, its intensity again diminishes by transmission and dispersion in the perpendicular plane, i.e., its image becomes longer in the first reflection and then wider in the second. Allowing for reflective dispersion, the final image is also inclined obliquely to the plane of the reflecting surface. Light diminished in this manner is apparently insufficient to penetrate the ocular media and excite the retina (or other detective apparatus). It will appear altogether extinguished as Guericke discovered in 1672.[14] We have here a case of diminishing reflection due to transmission, to which Mach observed that as Kepler discovered reflection without refraction (critical angle), so did Malus discover refraction (transmission) without reflection.

An interface between two transparent media is partially transmissive and partially reflective. Maximum transmission occurs at perpendicular incidence when no deviation at all exists. At this angle, the reflected light has its maximum angle of deviation; it returns on itself. To achieve equal separation between the transmitted and reflected

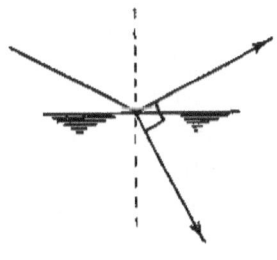

lights, their courses must be divided equally between maximum deviation and no deviation, i.e., between 0° and 180° which means a separating angle of 90°. When light is obliquely incident on an interface, the angle of transmission is a function of—among others—the refractive index of the medium (figure). Therefore, in order to find the angle of incidence that results in maximum separation of the reflected and refracted spectra, that is, a right angle, we employ Snell's formula: $n = V_1/V_2$ = sinα/sinβ=sinα/sin $(90 - \alpha)$ = sinα/cosα = tanα. This is also known as Brewster's law, and the angle α as the angle of polarization. Since n is always greater than 1, α must be greater than 45° (tan 45° = 1.) Polarization, as a function of velocities in and off media, appears to be but a particular case of refraction and reflection. (See also "Double Refraction").

Dealing with opaque metals, transmissivity is substituted by absorptivity, which varies of course from metal to metal and, hence, covariant with it is the angle of polarization—a fact also discovered by Brewster.

Phenomena of polarization lead us deeper and deeper into the structural nature of matter and its effect on light. The final short chapter carries the process one step further, for it deals with the encounter of light with completely opaque matter, which obscures direct inspection and renders the interpretation that much more speculative.

REFERENCES

1. Goldmann, H.: Stiles-Crawford-Effekt. Ophthalmologica 103; 225, 942.
2. Newton, I.: op. cit. *Opticks*, p. 372.
3. Young, T.: op. cit. *A Course of Lectures*, p. 541.
4. Hirschberg, J.: Geschichte der Augenheilkunde, in: Graefe-Saemish Handbuch, 2nd ed., vol. 12, Leipzig, Engleman, 1899, p. 156.
5. Mach E. The Principles of Physical Optics. Dover Publications; 1925: 34.
6. Sagnac, G.: L'ether lumineux demontre par l'effet du vent relatif d'ether dans un interferometre en rotation uniform. *Comp. Rend.* 157:708, 1913.
7. Pauli W. Theory of Relativity. New York, Pergamon Press 1958, p.7.
8. Kantor, W.: Direct First-order Experiment on the Propagation of Light from a Moving Source. *J. Opt. Soc. Am.* 52, 978, 1962.
9. Kantor, W.: Relativistic Propagation of Light. Lawrence, Kansas, Coronado, 1976.
10. Sommerfeld, A.: op. cit. Optics, p. 75.
11. Born. Op.cit. p. 407 and 262
12. Jenkins F. A., White H.E. Fundamentals of Optics. New York, McGraw-Hill; 1950 .
13. Malus. E. L.: Traite de. 1e. double Refraction. Mem. d. Sav. etrang. 2; 303-504, 1811.
14. Guericke, O.: Experimenta nova Magdeburgica. Amsterdam, Janssonium, 1672, p. 141.

Diffraction

"Lumen propagatur seu diffunditur non solum directe, refracte ac reflexe, sed quodam quarto modo diffracte." A proposition advanced by Francesco Maria Grimaldi (1618-1663): Light is propagated or diffused not only directly, by refraction and by reflection, but in still a fourth way—by diffraction.[1] The phenomenon, later termed by Newton as inflection, occurs when light is made incident on the edge of an opaque body. It is then deviated from its course both in direction of the body (into its shadow) and away from it (into the light that passed by), often leaving darkness in between. In this respect, the phenomenon differs little from that which ensues when light strikes the interface of a transparent body where it also deviates both in direction of the body (refraction) and off it (reflection), sometimes leaving darkness in between.

The results of Grimaldi's experiments vary somewhat according to whether the edge of one body is used, two bodies forming a slit between them, or whether light passes through one aperture or two successive ones. In general, light that is deflected from the body into the direct light forms fringes of various brightness and colors: a violet-blue tint on the side toward the body and a red farthest from it. Similar bands are produced by light deflected toward the body, except that the colors are complementary. This complementarity is similar to the one produced by refractive and reflective dispersion or the complementary colors of polarization. In addition, phenomena of interference ensue also by diffraction. The similarity of all these various effects suggests a common cause.

Diffraction implied to Grimaldi that the darkness in the fringes could perhaps be produced by the addition of lights and, furthermore, that light behaved like water bending around a breakwater. The analogy may be demonstrated by slowly introducing the handle of a knife, or similar body, into a stream of water issued by a faucet or hose. At first, the stream deviates toward the body, but as the body is advanced into the stream, or the latter's speed increased, the water deviates away from the body.

Isaac Newton goes into Grimaldi's experiments in some depth, [2] describing the influence of gravity on the course of light, its deviation, and its dispersion (figure).

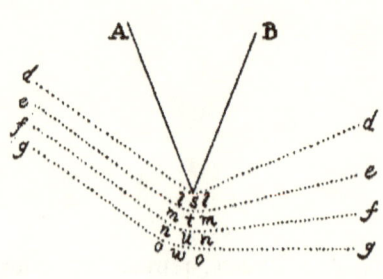

The rays of light that are in our air in their passage near the angles of bodies, whether transparent or opaque, are bent or inflected round those bodies as if they were attracted to them; and those rays which in their passage come nearest to the bodies are the most inflected, as if they were most attracted.

Dr. Young demonstrated diffraction by reflecting light unto a slit and then letting the light that passed through fall on two additional slits some distance apart (figure). The final image from these two slits revealed alternating areas of light and darkness,

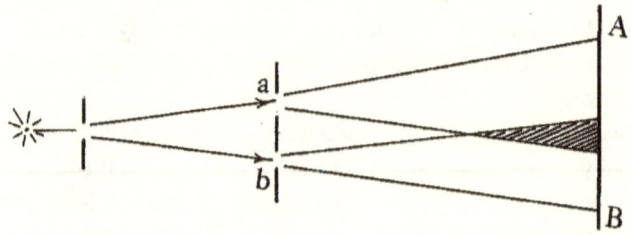

which led Young to believe that light must have destroyed itself where he expected some but saw none (compare to Fresnel's biprism experiment). The distance between a and b on the first screen, and A and B on the second is longer than if the rays went straight from the source to the screens; that is, light dilated. And as in the refracted or reflected spectrum, the violet was the most deviated.

It cannot be denied that by means of Young's experiments, accomplished with unusually clear philosophical insight and no mean experimental skill, and also without any great financial outlay, the science of optics experienced a very considerable advance. The prospective research student might observe from this that an important and apparently impenetrable truth is often really only very lightly obscured, and is readily disclosed when the seeker makes an attack in the right direction.[3]

On May 29, 1919, a group of scientists, observing a solar eclipse near the equator, discovered that light passing by the body of the sun was inflected from its course[4] (the light that deflected off this body was not visible in the sunlight). The discovery was then heralded in the professional and lay press as confirming an exiting and popular new hypothesis concerning the nature of matter, time, and light while those who still remembered Grimaldi were quietly puzzled. The figure here depicts Young's conception of inflection (diffraction) around a dark round object as he published in 1802.[5] All that is needed is to replace this object with the image of the sun or a black hole to see the correspondence, though some conservatives have objected.[6] Searching for a unified theory that could explain the largest number of phenomena with the smallest number of hypothetical assumptions may perhaps be helped by looking at the similarity of events in refraction, reflection, diffraction, and the bending of light near black holes.

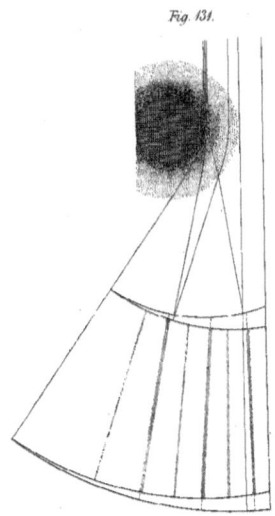

Fig. 131.

REFERENCES

1. Grimaldi, F. M.: Physico-Mathesis de lumine. Bologna, 1665.
2. Newton I. Mathematical Principles. Univ. of Calif. Pres. 1966, Vol. 1: 230
3. Mach, E.: op. cit. p. 275.
4. Dyston, F. W., Eddington, A. S., Davidson, C: Determination of the Deflection of Light by the Sun's gravitational Field. *Phil, transact.* 220; 291, 1919.
5. Young T. Of Colours by Inflection. In: Peacock G. Misc. Works. London; J. Murray. 1855. Vol. 1: 164-66
6. Ditchburn RW. Optokinetics (book review). *Optica Acta* 1982; 29:12

LINEAR MOTIONS
IN MOVING FRAMES

Linear Motions in Moving Frames

The advent of modern interplanetary spaceflight has brought to the fore practical issues of motions within moving frames of reference. When a rocket leaves the earth, its motions and speed are primarily referred to the moving earth. In interplanetary space, the motions may be referred to the sun or the star Deneb or Antares;[1] and when it nears its destination, be it the moon or a space station, the rocket's motions and positions are referred to this last frame. In between, the spacecraft is inertially guided by gyrocompasses.[2]

The resulting motions are calculated by computers based on old established Newtonian equations. "Thus the 'sputniks' constitute the first *experimental* proof of Newtonianism on a cosmic scale."[3] We argue in this volume that the motion of light in moving frames is essentially the same as the motion of material objects. It does not follow, of course, that light is identical to a material object, just as you and I are not identical because we may have an identical motion on the same airplane. As background, it thus seemed fitting to briefly refresh our memory of the history of these classical motions.

As is often the case in other issues in Western science, the need for a point or frame of reference when dealing with motion arose originally with the ancient Greek. The problem was closely associated with the very important question of whether the earth, along with its human beings, was the stable central frame of reference around which the firmament revolved, or did the earth move in reference to another point? The history of the conflict had been dealt with many times before, most recently in the erudite and most exhaustive treatise by Barbour;[4] we, therefore, select here only some ideas relevant to our thesis.

Aristotle

The Pythagoreans (about 530 BC) held that the earth with its counterearth moved around a vague central fire, perhaps the guardhouse of Zeus. Aristarchus (310-230 BC) actually theorized that the earth revolved around the sun on the circumference of a circle. On the other hand, geocentricity, and tacitly with it anthropocentricity, was advocated by Aristotle (384-322 BC) and later developed in great detail by Hipparchus (about 160-124 BC) and Ptolemy (about 127-151 AD).[5] Aristotle's ideas were of paramount importance not only because his was the greatest collection and systematization of knowledge until the Renaissance, upon which others then built, but also because it corresponded

with Scripture. Aristotle's position may be summarized in one paragraph:[6]

> That the centre of the earth is the goal of their movement is indicated by the fact that heavy bodies moving towards the earth do not move parallel but so as to make equal [right] angles, and thus to a single centre, that of the earth. It is clear, then, that the earth must be at the centre and immovable, not only for the reasons already given, but also because heavy bodies forcibly thrown quite straight upward return to the point from which they started, even if they are thrown to an infinite distance. From these considerations then it is clear that the earth does not move and does not lie elsewhere than at the centre.

Generally, Aristotle held as true five principles:

1. The earth is absolutely stationary.
2. The firmament moves in reference to the earth.
3. The earth is at the center, and its center is the center of the universe.
4. Heavy bodies move to the center of the earth in straight lines.
5. Heavy bodies when thrown upward return to the point on earth whence they started.

He also maintained elsewhere that motion was a process, an event, that required the continuous operation of a causative agent in order to keep it going. Reasoning by syllogisms, as was his way, all five principles were interdependent, one proven by the other.

When the geocentric position began to be seriously questioned early in the Renaissance, disproving any one of these principles was evidently deemed necessary and sufficient in order to support the opposing heliocentric doctrine. For instance, if it could be shown that heavy bodies did not move to the center of the earth in a straight line or returned to the position whence they were thrown upward, then Aristotle's view of the earth's centrality and immobility would be refuted. Persuasive observational means to contradict Aristotle were on hand after Copernicus and after Galileo when terrestrial experimental methods improved and gained credence.

Oresme

After Aristotle's works became available in Latin, new ideas about motions emerged, such as those expressed by William of Ockham (c. 1284-1349). According to him, all that needed to be said about motion was that from instance to instance a moving body had a different spatial relationship with some other body, a different stationary position. Every new effect required a cause, but motion was not a new effect since it was nothing except that the body rested successively in different places and, therefore, did not require a motive force as Aristotle taught.[7]

The next relevant author to our subject was Nicole Oresme (d. 1382), future bishop of Paris who wrote his *Le Livre du ciel et du monde* about 1377. Because of the Parisian condemnation of 1277, scholars had to tread Aristotle cautiously, for some of his views were approved by the church while others condemned, leading Oresme to express his opinions as if by someone else, a style later employed also by Galileo in his *Dialogue*. "Should someone say that the definition of local motion is to be transported to a different place with respect to some other body at rest, then, if no body were at rest, no body could move."[8] The decision as to which body is at rest and which one moved in reference to the other was arbitrary. Oresme illustrated his point by two isolated bodies:

> Let us suppose that *a* moves and *b* rests; then *a* and *b* would change their relative positions completely just as though *b* moved and *a* rested. In such a case, it would be impossible to explain why *a* should move rather than *b,* or vice versa, if it were true that motion implies a change of relative position with respect to some other body. Again, I am arguing especially about celestial motion. [9]

Later on he explained the following:

> In the same way, if a *man in the heavens* (moved and carried along by
> their daily motion) could see the earth distinctly and its mountains,
> valleys, rivers cities, and castles, it would appear to him that the
> earth was moving in daily motion, just as to us on earth it seems as
> though the heavens are moving.[10] (Emphasis added)

Concerning Aristotle's example of a stone thrown upward, Oresme agreed, but
did not accept it as evidence that the earth stood still.

> The case of an arrow or stone thrown up into the air, etc., one might
> say that the arrow shot upward is moved toward the east very rapidly
> with the air through which it passes, along with all the lower portion
> of the world which we have already defined and which moves with
> daily motion: for this reason the arrow falls back to the place from
> which it was shot into the air.

In general, when a body moves in one direction while the frame also moves, its
final course is determined by *compounding* the vectors of the two (figure).

> If they were not thus moved, *a* would go straight upward along the
> line *ab*; but because *b* is meanwhile drawn toward *c* by circular and
> daily motion, it appears that *a* describes the line *ac* as it ascends and
> that, therefore, the movement of *a* is compounded of rectilinear and
> circular motion.[10]

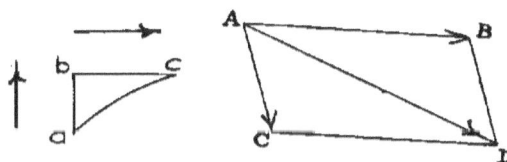

The rotation of the earth carried with it all objects at the same rate so that objects
thrown up fall back to the same position they have left. Oresme then used the example
of "a man in a ship moving rapidly eastward without his being aware of the movement,"
a case that later became classical when used by Galileo,[11] Newton,[12] or Einstein[13] in
order to illustrate an inertial system. "Inside the boat moved rapidly eastward, there

can be all kinds of movements—horizontal, criss-cross, upward, downward, in all directions—and they seem to be exactly the same as those when the ship is at rest." In support of pure mental ideas, Oresme promoted—as did Francis Bacon (1561-1626) after him—actual perceptual evidence in contrast to the then prevailing scholasticism: "I conclude, then, that it is impossible to demonstrate by any experience that the heavens have a daily motion and that the earth does not have the same." Unfortunately, Oresme's writings remained obscure until more recent times.

The principle discussed by Oresme has been later variously named: the law of inertia, inertial system of coordinates, a Galilean system, or according to Poincaré, the principle of relativity.[14]

> The principle of relativity, according to which the laws of physical phenomena should be the same, whether for an observer fixed, or for an observer carried along in a uniform movement of translation; so that we have not and could not have any means of discerning whether or not we are carried along in such a motion.

Albert Einstein, in his papers, adopted these words and definition for his own purposes, leading some to occasionally mistake this *principle* for his quite different *theory* of relativity. In his words,[15]

> We wish to elevate this assumption (whose contents shall subsequently be named "Principle of Relativity") to a precondition, and in addition import a seemingly incompatible precondition, that light in empty space always advances in a definite speed V, independent of the state of motion of the emitting body.

Copernicus

Nicolaus Copernicus (1473-1543) recognized early that the point of reference to which the motions of the heavens applied was the essential difference between his heliocentric concept and that of Aristotle and Ptolemy; was it the sun or was it the earth?

> For every apparent change in place occurs on account of the movement either of the thing seen or of the spectator, or on account of the necessarily unequal movement of both. For no movement is perceptible relatively to things moved equally in the same directions—I mean relatively to the thing seen and the spectator. [16]

He then brings as example the motions of a ship, similar to the one used by Oresme:

> And things are as when Aeneas said in Virgil: "We sail out of the harbor, and the land and the cities move away." As a matter of fact, when a ship floats on over a tranquil sea, all the things outside seem to the voyagers to be moving in a movement which is the image of their own, and they think on the contrary that they themselves and the things with them are at rest. So it can easily happen in the case of the movement of the earth that the whole world should be believed to be moving in a circle.

One of Aristotle and Ptolemy's argument against a moving earth was that objects on it would then fly away centrifugally, as Galileo later put it, "There now remains the objection founded upon which experience shows us, namely, that a swift whirling about has a faculty to extrude and disperse the matters adherent to the machine that turns around." To which Copernicus answered, "But why didn't he [Ptolemy] feel anxiety about the world [universe] instead, whose movement must necessarily be of greater velocity, the greater the heavens are than the earth?"

> Surely if this reasoning were tenable, the magnitude of the heavens would expand infinitely. For the farther the movement is borne upward by the vehement force, the faster will the movement be . . . and the immensity of the sky would increase with the increase in movement.

All the heavenly components would fly centrifugally away even faster than those on earth itself, for they are farther from the center of motion—an early reference to the modern belief in an expanding universe. [17] Aristotle's argument was not principally wrong, only by degree, for if the earth spun so fast as to exceed gravity, we all would indeed be propelled into outer space.

Which then is the immobile frame of reference to which motions relate? Oresme believed that all motion was referred to a local frame, citing Witelo (c. 1230-1280), "We do not perceive motion unless we notice that one body is in the process of assuming a different position relative to another." Though in an apparent attempt to please the church, he added that beyond this world may be an imaginary space infinite and motionless to which celestial motions could be related. Copernicus assigned the frame of reference to the immoveable fixed stars:[18] "Summus est stellarumfIxarum immobilis et omnia omnia continens et locans" (the highest [heavenly circle] is the fixed stars, which is immobile, contains all, and locates all).

However, in a broad philosophical (metaphysical) sense, Newton's skepticism seemed justified: "And therefore as it is possible, that in the remote regions of the fixed stars, or perhaps far beyond them, there may be some body at rest; but impossible to know . . . it follows that absolute rest cannot be determined from the position of bodies in our regions." Naturally, then, without an absolute point of reference, there could be no absolute motion. Though others, such as James Clerk Maxwell, in search of universal rather than local solutions, maintained that "there was at least one universe-wide reference frame: the all-pervading ether."[19, 20] The existence of this ether was, of course, set in serious doubt later by the negative results of Michelson's experiment.

At the end of the day, for practical purposes such as space travel, Ernst Mach's dictum holds true:[21] "The heaven of fixed stars is the only practical usable system of reference." But as Poincaré pointed out,[14] aberration constantly changes the position of the fixed stars so that the position of the earth in relation to them is basically a statistical average.

Kepler

A later scholar who also referred his views to Witelo and Aristotle was Johannes Kepler (1571-1630), particularly in his *Ad vitellionem paralipomena, quibus astronomiae pars optica* of 1604. So impressed was he with Aristotle that he translated verbatim the relevant Greek chapters into his own vernacular Swabian German, adding comments and objections. In his *Epitome astronomiae coperniacanae* (1618-1621), he said explicitly, "This book is designed to serve as a supplement to Aristotle's On the Heavens."

"Weight is nothing but the magnetic attraction of the earth." Kepler used the term *magnetic* to describe gravity, distinct from the modern term introduced by court physician William Gilbert in 1600 in his *De Magnete*. Therefore, falling stone or other heavy objects must aim to the center of the earth and not the center of the universe. To Aristotle's argument about a stone thrown upward, Kepler used a similar answer as Oresme's: the pull of the earth was distributed over its entire body; therefore, while rotating, it pulled around the stone in its flight so that it fell back to the place from whence it was thrown upward. In other words, the stone was synchronously moved along by the momentum imparted to it by the moving earth, and its return to its place of origin was therefore no argument against the earth's rotation.

> The reason people easily believe that the earth stands still is based on our point of view from the earth surface itself. Everyone must concede, however, that *if we were on the moon* (emphasis added) we would then similarly believe that the moon stands still, and construct astronomy from this point of reference. [22, 23]

In a figure in his *Epitome,* Kepler quantitatively traced the fall of an imaginary body from a most remote (*remotissimo*) place (not one thrown up from earth and not from the top of a tower on earth), dividing the distance on earth and the distance the stone traveled into fourteen units of time; the closer to the earth, the faster the stone's speed "because gravity is stronger from close distance."[24] While the body was thus accelerating, the earth rotated uniformly to the east, pulling the approaching body a bit with her so that it arrived on her surface east of its projected point of departure. Unlike Aristotle's view, its path was not perpendicular but had a quasi-spiral shape, and additionally, its fall ended on the earth's surface, not its center.

It is most positively quintessential for the reader looking at Kepler's drawing, and all similar ones, to be vigilant and always keep in mind that the circle on the page representing the moving earth and the line picturing the path of the falling object depict the circumstances as *seen from a stationary position outside the earth and the page,* such as from the moon, as Kepler wrote, or the "man in heaven," in Oresme's words. It is fundamentally imperative to pay constant attention to the position of the point of reference from which the motion under discussion is viewed. This is not easy because we and the page are in the same frame. The pictured circumstances are purely an intuitive product of the imagination—the view from the mind's eye—without any direct evidential correlates in physical reality whatsoever the actions are in our thoughts.

Carelessness in this regard, as we shall see, has led to some serious errors. Mach termed the procedure *gedankenexperiment* (thought experiment), for "our *imagination* is more *easily* and comfortably available than the *physical* facts"[25] (his emphasis). The method of thought experiments as opposed to real ones was favored, for instance, by Einstein in his arguments: "Thus with the help of certain imaginary physical experiments." And repeatedly, "Wir denken uns" (We imagine).[26, 27]

The first demonstration of the earth's motion, its spin, by experiments on the earth itself was accomplished by Jean Bernard Léon Foucault in 1851 by means of the inertia of a weight swinging at the end of a string that counteracted the earth's inertial motion.[28, 29] Today, this is done with gyroscopes, a name coined by Foucault. Since their history has not yet been fully recorded, it may perhaps be useful to offer here Kepler's ideas on the inertia (angular momentum) of a spinning top (figure),[30] particularly since, as Holton noticed, Kepler's style and Latin are so hard to decipher:

> *In all local motion, besides that which provides a static place, there must be in addition something else on which the moving object rests that causes it to remain in a state of inertia: for example, the centers in a lathe, a surface in a top, the air in flight, the waves in swimming, the Earth in walking. My question is: what in the case of the Earth causes it to remain in a state of rotational inertia?* (His italics)

Nearly the same thing that occurs in a children's top before it touches the surface. First, the entire globe of the Earth stays in its place. Its sections, however move into those preceding them, one after the other. Next, to get to specifics, the globe remains immobile according to rectilinear pull parallel to the axis, whereby the axis and poles are recognized, insofar as the primary motion is concerned. Moreover, the entire globe is subject to this motion due to the pull of the circumference between the poles. This motion rests in that state of inertia just as if the globe were to move fixated between the immobile centers of a lathe. All of this, naturally, is thus stated about the Earth just as it must be stated about the top, since it also rotates while spinning in the air.[31]

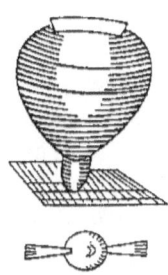

Benedetto may have thought along similar lines.[32] Elsewhere, Kepler again compared the earth's spin to the inertia of a child's top overcoming the resistance of its weight.[23] A recent comment on the subject (similar to Mach's) probably deserves repeating:

A little caution should be exercised when interpreting Foucault's experiment. The pendulum's plane of oscillation precesses (relative to a grid on the floor, say), and we infer that the earth rotates. But relative to what? Our theory tells us that the earth must be rotating relative to an inertial frame of reference. The stars in our neighborhood of the Milky Way provide a good approximation to such a frame. It would be unwarranted, however, to infer that the earth rotates—period. That inference would require a satisfactory definition of "absolute" rotation, and no operational definition, at least, is yet available.[33]

Galileo

The observational foundation to understanding the dynamics of terrestrial motions themselves, and in moving frames, was laid by Galileo Galilei of Pisa (1564-1642). Attempting to ascertain the nature of acceleration of a falling ball on an inclined plane, and thereby the force g of gravity, he discovered that when the plane turned to the horizontal and the force of gravity stopped, the ball persisted in a linear uniform motion. "But, upon the horizontal plane GH [figure] the body would maintain a uniform velocity equal to that which it had acquired at B after fall from A."[34] The phenomenon was described as "laziness of a dead body" by Kepler, and then termed principle of *inertia*, or the law of perseverance (*Beharrungsgesetz* in German).

Contrary to Aristotle and his followers, Galileo demonstrated experimentally that no force was necessary to keep a body either at rest or in uniform linear motion. Consequently, no distinction on the nature of forces and motions would be found between events occurring in a resting frame or one which is uniformly moving in a straight line. This state of affairs was sometimes termed the Galilean relativity principle corresponding to Oresme's ideas. To illustrate the circumstances, Galileo resorted in greater detail to Oresme's ship example:

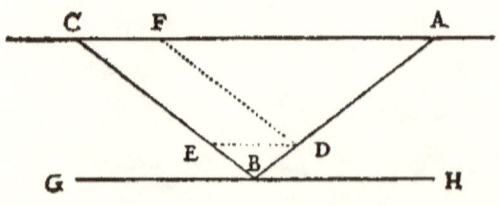

Then, the ship lying still, observe how those small winged animals fly with the like velocity towards all parts of the room; how the fishes swim indifferently towards all sides; and how the distilling drops all fall into the bottle placed underneath. And casting anything towards your friend, you need not throw it with more force one way than another, provided the distances are equal; and jumping broad, you will reach as far one way as another. Having observed all these particulars, though no man doubts that, so long as the vessel stands still, they ought to take place in this manner, make the ship move with what velocity you please, so long as the motion is uniform and not fluctuating this way and that. You shall not be able to discern the least alteration in all the forenamed effects, nor can you gather by any of them whether the ship moves or stands still.[11]

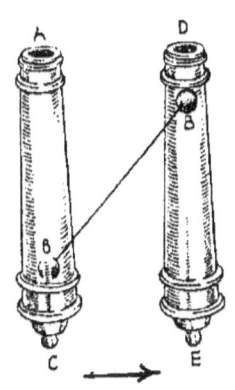

As to Aristotle's argument that the return to its origin of an upward thrown stone proved the immobility of the earth, Galileo answered with Oresme and Kepler that the moving earth simply carried with it both the body and the observer equal distances. The figure illustrated his point by having a cannon shoot up a ball vertically and while the ball moved up from point C to point A and the earth with the cannon moved from C to E, the ball also moved with it to point D out the muzzle.[35] Galileo then concludes, "For the real truth is that, as concerns these shots, the same ought exactly to befall as well in the motion as in the rest of the terrestrial globe." As he previously said, "The error becomes clearer in making the experiment in a ship with a missile shot upwards perpendicularly with the crossbow, which returns to the same place whether the ship does move or stand still."

Galileo took up Aristotle's hypothetical case of a heavy object falling from the top of a tower to the earth, which was itself rotating from. The tower moved in direction of the earth while the stone accelerated down, ending at the tower's base. "Now these points become more distant from the top of the tower in an ever-increasing proportion, and that is what makes its *straight motion along the side of the tower* [emphasis added] show itself to be always more and more rapid."[36, 37] In a later discussion on the difference between empirical physics and metaphysical mathematics, Galileo considered the hypothetical case of a ball "coming from the concave of the *moon*, which is so great a distance off. *If* the ball had participated in the earth's revolutions . . . that very venue which made it turn round before its descent will continue it in the same motion in its descending. And so far it is from not keeping pace with the motion of the earth, and from staying behind, that it is more *likely* to outrun it."

No matter how high his intellectual satisfaction may have been of answering Aristotle's metaphysical argument of a heavenly body descending to the center of the earth, Galileo, as always, was more inclined to trust practical sensory perceptions:[38] "But to us only one part of this motion is visible and observable, that is, the part of the *straight*, the other part of the circular being imperceptible to us, *because we are included in it*" (emphasis added).

Newton

Building upon Kepler's laws of planetary motions and Galileo's principles of terrestrial dynamics, Isaac Newton (1642-1727) quantified, formalized, and generalized them in his *Principia* of 1687. His Definition III concerns inertia: "The *vis insita* [inertia] or innate force of nature, is a power of resisting, by which every body, as much as in it lies, continues in its present state, whether it be of rest, or of moving uniformly forwards in a right line."[12] Mach thought that "his fifth corollary contains the only practical usable [probably approximate] *inertial system*."

"Corollary V: The motions of bodies in a given space are the same among themselves, whether that space is at rest, or moves uniformly forwards in a right line without any circular motion." And then Newton added Oresme's, or Galileo's, ship example: "A clear proof of this we have from the experiment of a ship; were all motions happen after the same manner, whether the ship is at rest, or is carried uniformly forwards in a right line."

Of interest to our thesis is also Newton's third law of motion: "Corollary I: A body, acted on by two forces simultaneously, will describe the diagonal of a parallelogram in the same time as it would describe the sides by those forces separately." An object moved in one direction, within a frame moving in another, will have a course and speed determined by the diagonal of the two (see figure under Oresme, right).[38]

These generally accepted universal laws of physical motions notwithstanding, Newton had earlier taken up Aristotle's old question of determining the earth's mobility by a stone falling to its center. He was perhaps stimulated by a report in the *Philosophical Transactions* of 1668 written by his Scottish friend James Gregory[39] in which all of Aristotle's arguments were, again, marshaled forth, including the path of a falling stone (from the 240-foot-tall tower of the Asinelli in Bononia) or of a cannonball shot up vertically. Newton details the problem in a letter to Robert Hook on

28 November 1679: "I shall communicate to you a fansy of my own about discovering the earth's diurna*l* motion."[40]

> Let A [figure] be a heavy body suspended in the Air & moving round with the earth so as perpetually to hang over ye same point thereof *B.* Then *imagin* this body *B* let fall & it's gravity will give it a new motion towards *ye center of ye Earth* without diminishing ye old one from west to east . . . it will not descend in ye perpendicular *AC,* but outrunning ye parts of ye earth will shoot *forward to ye east* of the perpendicular describing in it's fall a spiral line *ADEC,* quite contrary to ye opinion of ye vulgar who think that if ye earth moved, heavy bodies in falling would be outrun by its parts & *fall on the west side* of the perpendicular. (Emphasis added)

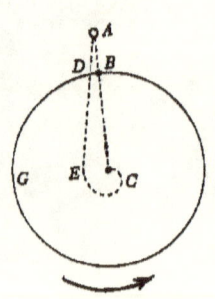

He then proposed an experiment where a "Pistol Bullet" was let fall to the bottom of a well in order to determine if it landed to the east of the predetermined perpendicular. (Newton mentioned it again in a letter to Halley on 27 May 1686.[41]) In his reply on December ninth, Hook agreed with Newton though disputing the exact path of the falling body:

> Tis certainly right & true soe far as concerns the falling of the body Let fall from a *great hight* to the Eastward of the perpendicular and not to the westward of it as most have hitherto Imagined. And in this opinion concurred Sr Christopher Wren Sr John Hoskins Mr Henshaw and most of those that were present at our meeting on Thursday Last.

It strikes one, perhaps, a bit melancholy to see Newton and his fellows in the Royal Society, having confessed allegiance to empirical Baconian principles, fail in this manner to rise above the Aristotelian scholastic agenda and seriously embark on an Icarian flight of metaphysical speculation, trying to trace the path of an object falling from a great height to the center of a vacant terrestrial globe. On the other hand, Newton may have recognized the difficulty when, two weeks later, in a letter to Hook on 13 December 1679, he wrote that "considering thus far the species of this curve . . . the thing being of no great moment." Some modern historians, for instance Korye[42] and Barbour, nevertheless believed that Newton's ideas had merit, whereas Galileo's were mistaken.[43] Amazingly, others have actually repeated the experiments by dropping objects into a well, with inconclusive results.[44]

Galileo quite clearly said that to an observer on the moving earth, objects fell perpendicularly, and from a tower, they landed at its base not to the east. His cannon illustration turned upside down fits the essence of Newton's proposed experiment: instead of Galileo's cannonball going through the muzzle straight upward without touching its sides, Newton's bullet goes through the well downward without touching its walls. Kepler had similar views, holding that objects descended to the earth perpendicularly not to the east, though this was no proof that the latter was immobile. (His example of an object falling from high in heaven to the *surface of the earth* differed from Newton's arrangement, for its starting point a priori did not participate in the earth's rotation and, therefore, did not have the angular momentum of a tower planted on it.)

Today, the influence of the earth's spin on objects hurled in the opposite direction *into outer space* (stationary in reference to the sun) cannot be totally ignored. The launch of a rocket will be assisted by this motion to a greater degree when done from southern Florida than from northern Maine, and more so when directed to the east than the west due to the faster spinning of the globe's surface to the east at the equator.

To summarize the work of Oresme, Copernicus, Galileo, and Newton concerning linear motion in a system, which moves uniformly in relation to another one, the adjacent figure is illustrative. There are two boxes, ships, or train cars, A and B, that, according to René Descartes, may also be designated as coordinate systems xy and x'y'.

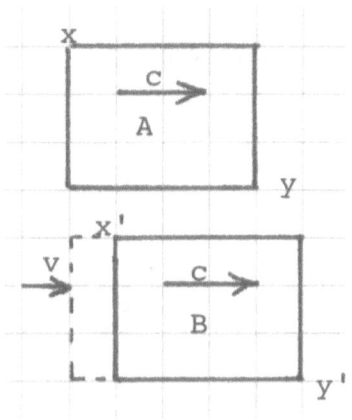

In system (ship) A, an object C is moving uniformly to the right with the speed c of 2 meters per second. In system B, a similar object is moving in the same direction with the same speed. As long as A and B are stationary in relation to one another and do not change their position on this page, the circumstances are equal. When system B is then moving to the right with uniform speed of 1 m/sec in relation to system A and in reference to the reader of this page, the speed of the object inside the system *as observed by someone inside the system* is also 2 m/sec—the same as observed in system A from inside that system. However, the speed c in system B is now 2 + 1 = 3 m/sec in reference to an observer in stationary system (ship) A and to the reader of this page. The speed v of system x'y' is added to the speed c of the object inside it.

When the system moves in opposite direction (from right to left), the speed is subtracted. When this speed is 2 m/sec (the speed of C), the object C appears to the reader and observer on system A to be standing still. In other directions, the compounded speed depends on the angle between the motions.

REFERENCES

1. Battin Richard H., *An Introduction to the mathematics and Methods of Astrodynamics.* Amer. Institute Aeronautics and Astronautics, Reston, VA, 1999, revised ed., pp. 623-59.

2. Gray Andrew, *A Treatise on Gyrostatics and Rotational Motion.* Dover Publications, New York, NY, 1959.

3. Koyre Alexandre, *Newtonian Studies.* Univ. of Chicago Press, Chicago, 1965, p. 54.

4. Barbour J. *Absolute or Relative Motion*, Cambridge Univ. Press, Cambridge, 1989.

5. Ptolemy Claudius, *The Almagest*, in: R. H. Maynard (ed.): *Great Books of the Western World.* Wm. Benton, Chicago, 1952, p. 12.

6. Aristoteles, *De caelo*, by J. L. Stocks. Clarendon Press, Oxford, 1922, p. 296b.

7. Crombie A. C., *Medieval and Early Modern Science vol. II*; Harvard Univ. Press, Cambridge, MA, 1967, p. 32.

8. Oresme Nicole, *Le Livre du ciel et du monde*, Univ. of Wisconsin Press, Madison, WI, 1968, p. 369.

9. Ibid. Ref. 8, p. 371.

10. Ibid. pp. 523-25.

11. Galileo Galilei, *Dialogue on the Great World systems.* G. de Santillana (transl.), Univ. of Chicago Press, 1953, p. 199.

12. Newton Isaac, *Principia*, Univ. of Calif. Press, Berkeley, 1966, pp. 2-17.

13. Einstein preferred a train rather than a ship to illustrate a moving frame.

14. Poincaré H. The Principles of Mathematical Physics. *The Monist* 1905; 15; 1-24.

15. Einstein A. Zur Elektrodynamik bewegter Körper. *Ann der Physik* 1905; 17: 891-921.

16. Copernicus Nicolaus, *On the Revolutions of the Heavenly Spheres*, in: R. H. Maynard (ed.): *Great Books of the Western World*, reference 5, pp. 518-20.

17. Gamow George, *One Two Three . . . Infinity*, Bantham Books, New York, NY, 1979, pp. 328-35.

18. Rossmann Fritz. *Nikolaus Kopernikus*, Erster Entwurf seines Weltsystems. Hermann Rinn, Munich, 1948, p. 12

19. Maxwell James C., "A Dynamical Theory of the Electromagnetic Field," 1865. *Phil Transact. Roy. Soc.* 155, 459-81. Also in: *Scientific Papers* (1890), Vol. 2.

20. Friedman Alan J. n, Carol C. Donley, *Einstein as Myth and Muse* Cambridge (1985), p. 49.

21. Mach Ernst, *The science of mechanics.* The Open Court Publishing Co., La Salle, IL, 1942, p. 295

22. Rossmann Fritz. Op.cit. Reference 18 p. 89.

23. Frisch Christian, *Joannis Kepleri Opera Omnia.* Heyder & Zimmer, Frankfurt A.M, 1868, vol. 7, pp. 748-9.

24. Op. cit. (1864) vol. 6, p. 182.

25. Mach Ernst, *Erkenntnis und Irrtum.* J. A. Barth, Leipzig, 1926, 5th ed., p. 187.

26. Einstein Albert, "Elektrodynamik bewegter Koerper," *Ann. d. Physik* (ser. 4), 1905; 17, 894-896.

27. Einstein Albert, *The Principle of Relativity.* Dover Publications, New York, NY, 1923; p. 40.

28. Maxwel J C. *Matter and Motion.* Dover Publications, New York, NY, 1952, p. 86.

29. Tobin W. The life and science of Leon Foucault. Cambridge University Press. 2003: 133.

30. Holton Gerald, *Thematic Origins of Scientific Thought.* Harvard Univ. Press, Cambridge, MA, 1988, revised ed., p. 53.

31. Caspar Max, *Johannes Kepler Gesammelte Werke.* C H Beck, Munich, 1953, vol. 7, p. 85.

32. Benedetti Giovanni Battista, *Diversarum Speculationum Mathematicarum, & Physicarum Liber* (Nicolai Beuilaquae, Turin, 1585) pp. 160-61 and 183.

33. Baierlein Ralph. *Newtonian Dynamics.* McGraw-Hill, New York, NY, 1983, p. 223.

34. Galileo Galilei. *Two New Sciences.* Dover Publications, (MacMillan, 1914) New York, NY, (nd), p. 216.

35. Reference 11, p. 189.

36. Reference 11, p. 179.

37. Galileo Galilei, *Dialogue Concerning the Two Chief World Systems.* Univ. of Calif. Press, Berkeley, 1953, p. 165.

38. Reference 11, p. 259.

39. Gregory Jacob [sic!], "An account of a controversy," *Phil. Trans. Royal Soc.* 36, 693-98; 1668.

40. Trumbull H. W. *The Correspondence of Isaac Newton.* Cambridge Univ. Press, 1959-1981, p. 301.

41. Op. cit. p. 433.

42. Korye Alexandre, "A documentary history of the problem of fall from Kepler to Newton," *Trans. Amer. Philosoph. Soc.* 45 (4), pp. 329-95 ; 1955.

43. Carr Edward H., *What is history.* Random House, New York, NY, 1961, p. 24

44. Armitage Angus, "The deviation of falling bodies," 1947; *Ann. of Sci.* 5, 342-351.

OPTOKINEMATICS

Optokinematics

Speeds of over one billion kilometers per hour are hard to measure on this small earth we love and with the technology with which we pride ourselves. It is easier done in the vast space of the solar system where the scale better fits the task—no one will want to measure the distance from London to Peking with a yardstick or time a rocket with Grandfather's pendulum clock. Direct measurements from nature are also preferable to human modifications of natural events (experiments) because the latter, as a matter of course, introduce foreign elements whose effect on the outcome is often unforeseen. For instance, light traveling some length in a straight path may seem equivalent to its repetitive passage by reflection along a shorter path, but then we assume that reflection does not alter the light so that the validity of our results comes to depend upon the validity of the often unspelled assumption and are that much less certain.

In the following chapter, we shall look into the motion of light in free space in an attempt to gain some insight into the nature of its motion. First, we study the motion in space as seen by an observer (Roemer) who moved in the line of light, then the effect measured by an observer (Bradley) who moved perpendicular to the direction of light's motion, and finally, analyze in detail a terrestrial experiment that combined light's motion in both directions (Michelson). These three events occurred in the seventeenth, eighteenth, and nineteenth centuries are fairly well-known and easier to inspect than the recent past, obscured as it is by a plethora of untried data. In each case, before treating of the motions of light itself, we glance at analogues from mechanical motions solely in order to ease the comparison and obviate any differences or similarities.

Again, when we consider motion of any kind, we ought to be constantly aware of the frame of reference. A line drawn to represent a motion usually means two things: first, that the motion is in reference to a stationary page, and secondly, that we are outside the line and are stationary with the page. When we draw a circle and say it represents the earth's orbit around the sun, we mean that we look upon earth from outside the earth and are stationary with the frame of reference—the sun.

Speed in Line of Sight

ROEMER'S MEASUREMENT

By measuring the periods of Jupiter's satellites, Roemer determined that the speed of light was higher when approaching and lower when receding from Jupiter than when at a steady distance from it.

Strictly for the sake of illustrating motions, we employ the example of balls—familiar to almost everyone since childhood—not implying thereby, in any way at all, that light is indeed a corpuscular body, a photon, or motion of some ethereal medium. Take a stationary platform J (figure) whence balls are thrown toward E with regular periodicity of, say, every four hours (or minutes or any other unit). A man stationed at E receives a ball every four hours, say, at two, six, ten o'clock, and so forth. This man then moves to E', a distance (d) of 2.5 km farther from E. Stationed there, he may expect to receive a ball at six o'clock, but instead receives it at six thirty. Since the balls leave the platform at a fixed time and rate, the time difference between arrivals at the second and first position—six thirty versus six—means that the second ball took thirty minutes to cover 2.5 km and, thus, had the velocity of $c = d/t =$ 2.5/0.5 = 5 km/hr.

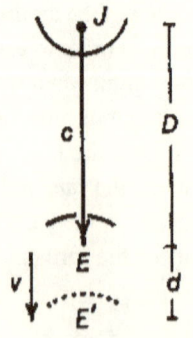

Instead of being stationary at E, a distance D (say 10 km) from J, take the man to be moving away from J at a uniform speed of $v = 1$ km/hr while the ball is leaving J. By the time the ball, which left J at time $0°°$, reaches position E, E itself has moved farther; the question is at what time will the man receive the ball? We designate with d the unknown distance from E that the man covers until he receives the ball at E', the time elapsed since leaving E until the reception is d/v. And this time equals the time it takes the ball to arrive from J: $d/v = (D + d)/c$. By transposition, we obtain the following: $dc = (D + d)v = Dv + dv$; $Dv = dc - dv = d(c - v)$; $D/(c - v) = d/v$.

We now look at the value of d/v we started with and arrive at the equation $(D + d)/c = D/(c - v)$, which means that the time it takes to cover the distance to reception at E' when E' is stationary equals the time it takes to receive the ball at a receding position E with the ball moving at the lower speed of c - v, or in reference to a man receding at speed v from a ball of speed c, the speed of the ball is lower than when he was stationary. The ball that left J at $0°°$ o'clock will reach E later, and the time period between balls will be longer. The opposite occurs when the man approaches the platform; namely, the periods shorten.

Replacing the letters with their numerical values of our example we have $D = 10$; $c = 5$, and $(c - v) = 5 - 1 = 4$ km/hr. It took two hours (10/5) for the ball to reach the stationary man at E. When he was receding, it took $D/(c - v) = 10/4 = 2.5$ hours. A ball that left J at $0°°$ hour reaches stationary E at two o'clock. The next ball leaves at four o'clock but reaches a receding E at six thirty; that is, the period increased by thirty minutes. (The simple numbers in our examples are of course for illustration only and are not at all proportional to the real distances, times, or speeds [Jupiter to earth or Io's periods]).

These all are empirical facts established by actual experience and not yet contradicted; when the time periods measured from a receding system increase compared to a stationary system, it means that the velocity in reference to the moving system is diminished, or in general, velocities in inertial frames of reference are not universally constant.

Between 1672 and 1676, the Danish astronomer Ole Roemer (1644-1710) —son-in-law of Dr. Erasmus Bartholin and, later, mayor of Copenhagen—worked in Paris under the patronage and with the help of the Italian astronomer Gian Domenico Cassini (1625-1712). The two engaged in measuring the periods of time it takes the Jovian satellite Io to revolve around its planet with the aim of synchronizing clocks around the world according to the positions of celestial bodies (establishing ephemerides)—of particular import to navigators at sea. In August of 1675, Cassini reported to the Academy in Paris some irregularities in the periods of Io, and in September 1676, Roemer reported his conclusion that these irregularities were caused by the different times it takes light from Jupiter to arrive to earth positioned at different distances from it.[1, 2]

Roemer actually discovered two facts. First, he found that the duration of Io's eclipses (the beginning or end of the period) as measured from earth when it was nearest Io occurred earlier on the clock than when the earth was farthest away. The difference was about twenty-two minutes; therefore, Roemer concluded, light must have a certain velocity, for it took time to cover the distance of the diameter of the earth's orbit around the sun, the same circumstances as exemplified above by a ball moving between two stationary positions some distance apart. The actual speed of light was later calculated from the time data and distances to the sun.

The second fact Roemer found was that each of Io's periods was longer when earth was in the process of receding from Jupiter and shorter when earth was on the approach than when it was at a fairly constant distance from the light source. This second measurement is hardly ever mentioned in the relevant literature and may have been forgotten. The difference between one period measured from a fairly stationary position compared to the same period measured from a receding or approaching position was quite small, but became obvious when Roemer added forty periods on approach compared to forty periods measured at rest or compared to forty periods on recession. As translated in the *Philosophical Transactions of the Royal Society* in 1677, it read,

> For, as M. Roemer had examin'd the thing more nearly, he found, that what was not sensible in two revolutions, became very considerable in many taken together, and that, for example, forty revolutions observed at the side F, might be sensibly shorter ["plus courtes" in the original French] than forty others observed in any place of the Zodiack where Jupiter may be met with.

Furthermore, the loss of time on approach equaled the gain on recession, and the mean of the two equaled the period measured from a stationary position. Roemer's data concerning light thus appear perfectly in accord with data derived from the general treatment of motion—the motion of anything at all. Again, Roemer discovered basically *two* things:

1. The *begin and end* of the period occurred at a later time when measured at point E (figure), a long *distance* from Jupiter, than when measured at H, a shorter *distance* from the planet. Therefore, it took light some time to cover the diameter of the earth's orbit around the sun.
2. At equidistance from Jupiter, the *length* of the period was greater when the *direction* of the earth was away from the planet from point L to K and shortest of all when the earth's *direction* was toward Jupiter at position F.

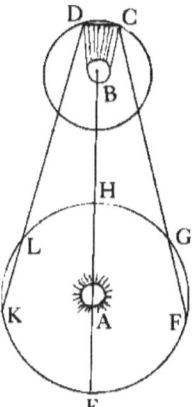

Ernst Mach [3] compared the revolutions of the satellite to the revolving sails of the windmill; their light is slower to reach a receding observer, and their revolutions therefore appear longer. The opposite occurs on approach when the speed of light in reference to the observer increases and the wings appear to rotate faster:

> In order to make Roemer's ideas quite clear, let us think of the revolving sails of a wind-mill. At a constant distance from an observer the revolution of the sails appear to be just as quick as it actually is. If, however, the observer moves very quickly away, the revolution must appear slower, because the light from each successive position reaches him later. The period of revolution apparently depends upon the *relative velocity* (emphasis added) with regards to the observer. The principle thus expressed differs from the well-known Doppler principle only in its application.

The rotating wings of the windmill may be replaced by the rotating hands of a clock. To a receding observer, they move slower. If, however, he believes that they should move just as fast as when he was standing still, he would have to conclude that time retarded or distance shrunk.

The procedures of timing the durations (periods) of Io's orbits brings us back to the problem of simultaneity and synchrony, which will be touched upon when dealing with the definition of time in the last section. The subject attained some prominence after Henri Poincaré mentioned that complications may ensue when one clock is in motion relative to another; and then, when Lorentz supposed that the speed of light between moving inertial frames was forever constant, they indeed materialized.[4]

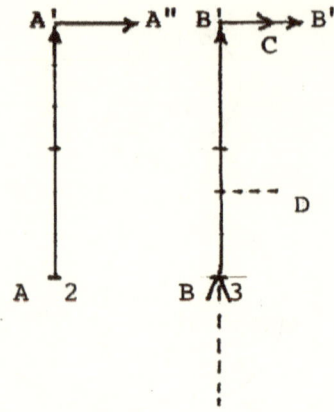

Let a signal from point A be sent to point A' 2 miles away at speed of 2m/hr. In one hour, a clock's hand moved from A' to A". When the signal is sent from point B, which is moving toward B' at speed 1m/hr, the signal's total speed will be 3 m/hr (speed of signal + speed of point B). It arrives at B' sooner; that is, the clock's hand only moved 2/3 of an hour to point C. However, if the signal (say, of light) from B is regarded to be the same as from A (2 m/hr), i.e., unaffected by the speed of its source, and the clock moved from B' to B", it would seem in this system that time dilated (from C to B"). Or else the signal must have originated at point D closer to B'; that is, the distance shrunk. Simultaneous events in A' and B' cannot be determined because the hands of their clocks seem to move at different speeds. Therefore, time (the rate of clocks) was said to be a local affair, different for each moving frame (or moving signal), as Poincaré first deduced.

Furthermore, the faster the signal's speed, the greater this discrepancy between the actual measurements, and the calculation that assumes that the signal's speed is unaffected by its source; that is, time dilated more or distances shrunk more. When the speed of the source at point B approaches the speed of light, time almost stops or distance almost vanishes. That is, the speed of a light is the maximum achievable speed of any kind—nothing can move faster.

Roemer's findings have been more recently reproduced with modern instruments,[5, 6] but the enormous velocity of light nevertheless does encumber easy access to accurate data. When the velocity of light relative to a stationary earth is taken to be 300,000 km/sec, it is about 300,030 km/sec when earth approaches Jupiter at maximum speed. When the time is measured over a course ten kilometers long, it traverses it in 1/30,000 parts of a second in the first instance and in 1/30,003 parts of a second in the second, a difference of 0,00 000 003 000 33 parts of a second. There is no readily available clock to measure such magnitudes, and moreover, there is no way to determine on this order of precision when the eclipse of Io begins and when it ends. This is where statistics become useful.

Dominique François Arago (1786-1853) understood Newton's theory of refraction to mean that the degree of refraction of the corpuscular rays varied with their speed of impact on the medium; the prism's glass in this instance. Accordingly, he reported to the French Academy in 1818 (not published until 1853) the results of numerous measurements of the prismatic refractive deviation of light from stars when the earth was on the approach and when it was receding from them, expecting the angles to differ. Lo and behold, "in manifest contradiction to the Newtonian theory of refraction," he found "the rays of all stars are subject to  the same deviations."[8, 9] Arago also measured with these primitive means the prismatic deviation of lights from sources with various intensities with the same null results; "the luminous rays issued from diverse stars, the sun, the moon, the planets and terrestrial lights, suffer the same deviation."

Considering the dim lights from the stars and the minute deviations of the spectra, Whittaker's remarked,[10] "The accuracy of Arago's experiment can scarcely have been such as to demonstrate absolutely his results." The change in refraction that occurs with motion toward and away from celestial bodies was firmly established only some years later by dint of Doppler's theory and Zöllner's actual measurements with a sophisticated spectroscope. In Larmor's words, "Arago worked with star light for which the Doppler effect due to relative motion would make a real difference, excessively minute however and beyond his observational means."[11]

James Clerk Maxwell conducted two similar sorts of experiments relating to the effect of motions on the angle of refraction of light.[12] First he tried, similar to Arago, to detect the effect of light coming from a receding (or approaching) star on the deviation of absorption lines in a prismatic spectrum. He expected such effect (redshift) since "if V_0 is the velocity of propagation of light in the luminiferous medium, and v_o is the velocity of the earth, $V = V_0 - v_o$, and the wave-length will be increased by a fraction of itself." He detected no such variation. He also carefully noted the decisive, but often overlooked, difference between observations through a succession of simple prisms and an achromatic arrangement:

> If instead of a spectroscope, an achromatic prism were used, which produces an equal deviation on rays of different periods, no difference between the light of different stars could be detected, as the only difference which could exist is that of their period.

In conclusion, Maxwell explained the negative results of the effect of motion on refraction (as Patrick Wilson, MD, already reasoned in 1782) as arising from the cancellation of the refractive deviation on entering the prism by its equal but contrary deviation on exit:[13] "Since the deviation of light by the prism depends entirely on the retardation of the rays within the glass, no effect of the earth's motion on the refrangibility of light is to be expected."

The second sort of Maxwell's experiments aimed at testing the change in the speed of light, measured by its deviation through a prism, by passage through a medium (prism) moving with the earth in different directions as compared to a stationary one. Here, however, both the *source of light and the observer were on the same moving platform*—the earth.

> The essentials of this experiment are *entirely terrestrial* [emphasis added], and independent of the source of light, and depend only on the relative motion of the prism and the luminiferous medium, and on the direction in which the ray passes through the prism.
>
> If the deviation of the rays in passing through the prisms from east to west differs from that produced during their passage from west to east, the image of the vertical spider-line formed by the rays which have traversed the prisms twice will not coincide with the intersection of the spider-lines as before . . . I have tried this experiment at various times of the year since the year 1864, and have never detected the slightest effect due to the earth's motion.[12]

Maxwell is quoted[14] to have said, "It [relative motion] cannot be determined by spectroscopic observations with our present instruments, and it need not be considered in the discussion of our observations." In a letter published a couple of months after his death,[15] Maxwell expressed his belief that the negative results were due to inadequate delicacy of the measuring procedure:

> And in the terrestrial methods of determining the velocity, the light comes back along the same path again, so that the velocity of the earth with respect to the ether would alter the time of the double passage by a quantity depending on the square of the ratio of the earth's velocity to that of light, and this is quite too small to be observed.

The idea was that the motion of the earth would alter the motion of the light, light that was emitted and observed on this same moving earth. Reasoning mathematically, he expected light passing through a receding prism to result in a less refracted spectrum (or spectral lines) than the spectrum from a stationary prism. And whereas Maxwell may have been mathematically reasonable, the experiment could not have possibly succeeded when the light source and the prisms were both on the same uniformly moving earth. That much had already been established kinetically by Oresme, Galileo, and Newton and their ship's cabin example; except, as we shall see, the principle was apparently not deemed relevant to optical phenomena occurring within an ethereal ether. The positive change in the position of absorption lines when an approaching or receding light source and the observer were both on earth itself was actually first demonstrated only in 1901[16] without any connection whatsoever to the relative motion of the earth. The final observational proof of Doppler's theory in reference to heavenly bodies was furnished only in 1869 by Friedrich Zöllner (1834-1882) with his ingeniously constructed very sensitive reversion spectroscope.

Ten years after his failed prisms' experiments, Maxwell said,[17]

> If it were possible to determine the velocity of light by observing the time it takes to travel between one station and another on the earth's surface, we might, by comparing the observed velocities in opposite directions, determine the velocity of the aether with respect to these terrestrial stations. All methods, however, by which it is practicable to determine the velocity of light from terrestrial experiments depend on the measurement of the time required for the double journey from one station to the other and back again, and the increase of this time on account of a relative velocity of the aether equal to that of the earth in its orbit would be only about one hundred millionth part of the whole time of transmission, and would therefore be quite insensible . . .

> The only practicable method of determining directly the relative velocity of the ether with respect to the solar system [including the earth] is to compare the values of the velocity of light deduced from the observation of Jupiter's satellites when Jupiter is seen from the earth at nearly opposite points of the ecliptic.

Maxwell's fundamental presupposition was that the velocity of light in opposite directions to the earth's movement would not be equal. He was an ardent believer in the hypothesis that light was a motion in and of the ether. In addition, he was possibly ignorant of Roemer's observations of the Jovian satellites, whose original time data were not published until the twentieth century,[2] or of the many tables of the satellite's periods already available at the time;[18] for as he said, if these values were available, the value of the earth's motion referred to the sun or to a stationary ether could be determined. Alas, he continued, "I am not an astronomer."[19]

Roemer's conclusions were acceptable to some—such as Huygens, whose theories and preconceptions it well fitted—but were opposed by others, such as the conservative Cassini, who even doubted Copernicus's concept of the solar system. Roemer's observations were reproducible, but the fact that light had a certain velocity was not generally accepted until confirmed in another manner by Bradley.

REFERENCES

1. Roemer, O.: Demonstration touchant le Mouvement de la lumiere trouve Par. M. Roemer. *Mem. de Fac. Roy. de Sc.* 1666-1699, Paris 1730, p. 575.

2. Cohen, I. B.: Roemer and the first determination of the velocity of light (1676). *Isis* 31; 327-79, 1940. Also: New York, Burndy Library, 1944.

3. Mach Ernst, *The principles of physical optics*, 1926; Dover Publications, New York, (n.d.), 23.

4. Einstein A. Relativity, The Special and the General Theory (1916). New York; Crown publishers, 1961: 17-29.

5. Goldstein S. J., Trasco J. D., "On the velocity of light three centuries ago," *The astronomical journal*, cxxviii (1973), 122-25.

6. Shea J. H. Ole Romer, the speed of light, the apparent period of Io, the Doppler effect, and the dynamics of Earth and Jupiter. *Am. J. Physics.* 1998; 66(7): 561-69.

7. Jammer M. Concepts of Simultaneity. Johns Hopkins Univ. Press, Baltimore, MD, 2006.

8. Arago DF. Memoire sur la vitesse de la lumiere. *Comptes Rendus* 1853; 36: 38-49.

9. Stachel J. Fresnel's (Dragging) Coefficient as a Challenge to 19th Century Optics of Moving Bodies. In: Einstein Studies. Birkhauser Boston, 2005; vol. 11: 1-13.

10. Whittaker E. A History of the Theories of Aether and Electricity.(2nd ed.) Harper & Brothers, New York, 1960: 109.

11. Larmor J. Aether and Matter. Cambridge University Press, 1900, p. 7

12. Maxwell, James Clerk, "On the influence of the motions of the heavenly bodies on the index of refraction of light," *Phil Transact Roy Soc*, clviii (1868) :532-535.

13. Wilson P. An Experiment proposed for determining, by the aberration of the fixed Stars, whether the Rays of Light, in pervading different media, change their velocity according to the Law which results from Sir Isaac Newton's Ideas concerning the Cause of Refraction. *Philosoph. Transact Roy Soc* London; 1782; 72: 58-70.

14. Zöllner Johann Karl Friedrich, "Ueber ein neues Spktroskop nebst Beitraegen zur Spektralanalyse der Gestirne," *Annalen der Physik*, cxxxviii (1869), 32-45

15. Maxwell, James Clerk, On the possible mode of detecting a motion of the solar system through the luminiferous ether. letter, *Nature*, xxi, (1880), 315. Also: *Proc. Roy. Soc.* 1880;30:1018.

16. Belopolsky A., "On an apparatus for the laboratory demonstration of the Doppler -Fizeau principle," *Astrophysical Journal,* xiii, (1901), 15-24.

17. Maxwell, J. C. "Ether," in *Encyclopaedia Britannica*, (1878), vol. 8: 568-72.

18. Denham Wm Observations of the eclipses of Jupiter's Satellites. *Phil. Transact.* (402) 1728: 415-428.

19. Maxwell JC. Clerk Maxwell and the Michelson Experiment. *Nature* 1930; 125: 566-67.

Speed in Perpendicular Direction

BRADLEY'S OBSERVATION

*By measuring the angle which light from a
star forms with an observer moving perpendicular to
it Bradley determined that the speed of light was
higher when approaching the vertical and lower when
receding from it than when positioned directly
under it.*

JAMES BRADLEY

From a stationary platform (J), a ball (or a rain of successive balls) is thrown into a stationary vessel or tube at E with the speed of c; the ball reaches the bottom of the tube without striking the walls. Now let the tube move to the right, with velocity v referred to J, at a right angle to the direction of the ball's motion. As the ball moves through the tube, the tube's left wall moves to the right and strikes it. In order to avoid this, the tube must be turned at a certain angle α to the right in direction of its motion. The tangent of this angle is given by the ratio v/c; and when velocity v and the angle α are known, we arrive at $c = v/\tan\alpha$, i.e., the velocity of the ball is a function of the angle and the velocity of the tube in reference to the same frame.

In reference to the angular position of the tube, the ball appears to arrive from J' to the right of J, i.e., in direction of the tube's motion. This is a fairly common experience of hunters or soldiers; when desiring to strike a perpendicularly moving object, the aim must be in front of it. (The motion of E referred to J equals that of J referred to E.)

Per unit time, the ball moves a distance D at velocity c, and the tube a distance d at velocity v. According to Pythagoras, the result of moving these distances is $D' = \sqrt{D^2 + d^2}$, and the velocity compared to a stationary tube or platform is $c' = \sqrt{c^2 + v^2}$, which is greater than c, or $c/c' = \cos\alpha$; $c' = c/\cos\alpha > c$ (because cos < 1). A ball moving at a given velocity c in reference to a stationary tube will be caused to move faster in reference to a tube in motion toward the perpendicular to the direction of the ball's motion, and this fact may be determined by noting whether the tube is turned at an angle compared to its position when stationary. The tube's motion causes an effect manifested by increased velocity and changed angle.

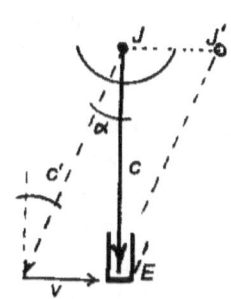

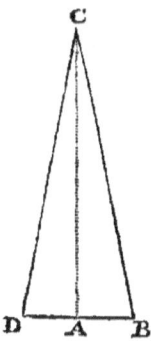

In the early part of the eighteenth century, a controversy was alive as to whether the fixed stars exhibited a parallax observable from earth. James Bradley (1693-1762), later Astronomer Royal, and his rich friend Samuel Molyneux set out to investigate the problem in the latter's home in Kew by London. They aimed a telescope at a bright star in the constellation Dragon, Gamma Draconis, that, in that latitude, was almost straight overhead—thus to avoid atmospheric refraction and aid accurate positioning of the telescope relative to a plumb line. As Bradley reported to the Royal Society in 1728, these observations revealed that all stars overhead seemed to move in direction of the earth's motion around the sun and, during the course of one year, completed a full circle whose diameter subtended about 40". Stars near the horizon seemed to be moving in a straight line in direction of the earth's motion, and the length of this line also subtended an angle of about 40" at the eye.[1, 2, 3]

When the earth in its annual orbit went from B to A (figure, form Bradley's original paper), Bradley had to change the direction of his telescope from straight upward (AC) to a little forward (BC) in order to see the star overhead (C), (from angle D to d_i in the figure), and when orbiting the other way, the tilt of the telescope was also reversed, tilted in reverse at the same angle.

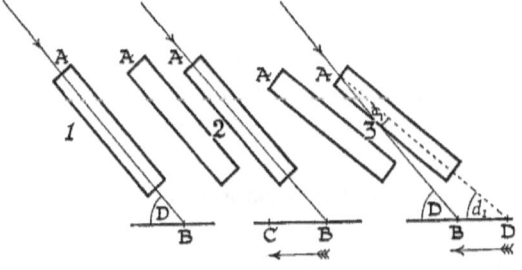

Having pondered the phenomenon for some time, Bradley concluded that the angular *aberration* in the position of the star was an effect caused by the compounding of the motion of the observer on earth moving in one direction (B to A) (in reference to the fairly stationary extraterrestrial firmament) with the motion of the light (C to A) moving almost perpendicular to this observer. The value of the earth's velocity and the angle of aberration being known, Bradley deduced the velocity of light ($c = v/\tan\alpha$), and his result concurred very well with the then only available other data obtained by Roemer's method.

The fact of aberration means that the velocity of light (c') referred to a moving earth ($\sqrt{c^2 + v^2}$; or c/cosα) is greater than the velocity of light referred to a stationary earth. Just as the speed of light varied with the speed of the observer in Roemer's measurements so did it in Bradley's. Accordingly so far, in general, the speed of light evidently follows the general principles of all motions as established by Galileo and Newton.

Later, in the nineteenth century, the phenomenon of aberration stuck as a thorn in the body of Huygens's and Young's prevailing wave theory, perhaps due also to the language of Bradley's original report of its discovery in 1729, using Newton's term "particles of light." The complete faith in the actual existence of a material ethereal ether was then the firm basis for interpreting optical phenomena and, as we shall see, the prime cause of the ensuing problems. Embarrassing as it may look today, Charles Biot[10] actually calculated the force which the molecules of this substance called ether attracted each other, and found it to be in the inverse ratio of the sixth power of their distance.

Though Bradley's measurements may be easily understood according to classical principles of kinetics, they gave rise to innumerable other explanations because they stood in contrast to subsequent theories about the nature of light. Doppler, for instance, detailed five different attempts at explanation.[4] Sir Oliver Lodge saw it in relation to the ether,[5] and Einstein saw it in his own relativity way.[6]

Roger J. Boscovich (1711-1787), "as early as 1766, conceived of an experiment which, if it had been carried out at that time, would have become one of the corner-stones of modern physics, and would have accelerated its development by many years: namely, *to measure the aberration of starlight by means of a telescope filled with water,* in order to discover whether or not the velocity of light in air and water remains the same, or differs by measurable amounts."[7] At the time, Boscovich ordered for this purpose a twin telescope—one filled with water, the other, with air. Unfortunately, he lost his job in Milan before actually performing the experiment. According to Newton's concepts, the speed of light increased when passing through water, and therefore, the angle of aberration was thought to differ if measured by a telescope filled with water. Dr. Patrick Wilson, on the other hand, argued that, though the speed of light in denser media (water) is higher, the angle will not thereby change because the angular refraction of the light entering the telescope will be reversed on its exit. In addition, as noted earlier, Arago later thought to have proven by his experiments that the refraction by a glass prism was not affected by the speed of light impinging on it, contrary to Newton.

Arago communicated the problem to his friend August Jean Fresnel (1788-1827) who theorized that the ether, which constituted light, was moved and was dragged along with the moving medium (water). Therefore, no difference would be found whether the telescope was filled with air or water or whether the refraction was measured when the earth approached or receded from the light emitting star.[8, 9]

In the preface to his book *On the Undulatory Theory of Optics*, Astronomer Royal George Biddell Airy said,[11] "The Undulatory Theory of Optics is presented to the reader as having the same claims to his attention as the Theory of Gravitation: namely, that it is certainly true." Airy was stimulated to experimentally test Fresnel's dragging theory by publications in 1867 by Hoek and Klinkerfues:[12] "Professor Klikerfues had computed that the effect of the 8-inch column of water and of a prism in the interior of the telescope would be to increase the coefficient of Aberration by eight seconds of arc." He did not mention the prior communications by Boscovich or Wilson, probably inadvertently, for at a time when university libraries were small, and Internet search engines in the far distance, older publications were often missed. For example, though mentioned in private letters, I cannot find Kepler's name in any of Newton's publications, and his very extensive library held none of his books.

Airy reasoned that since the angle of aberration was due to the relative velocities of light and earth, it should be possible to change the angle by changing the velocity of light, and this could simply be done along the last few inches of its already altered direction. It was thought possible to bypass the actual cause of aberration (the velocity of earth relative to the star) by changing the effect (the resulting final velocity of light) and, thereby, observe a change in the other effect (the angle of aberration). Some twenty years earlier, Foucault and Fizeau discovered that the velocity of light in water was lower than in air.[13] This was celebrated as confirming Dr. Young's undulatory theory as opposed to Newton's corpuscular one, ignoring the fact that the velocity of sound waves, which Dr. Young used as analogy to form his hypothesis, was higher in denser media.

"With the sanction of the Government," Airy aimed a telescope, 35.3 inches long, at the same star—Gamma Draconis, which Bradley observed—millions and millions of miles away and measured the angle of aberration. He then filled the telescope with water and measured the angle again. Lo and behold, the angle did not change. "True Aberration is the same as the Received Aberration."[14]

It is difficult to overestimate the impact of Airy's experiment on developments in optics. For instance, as Michelson put it in his 1887 paper with Morley,

> It may be noticed that most writers admit the sufficiency of the explanation according to the emission theory of light; while in fact the difficulty is even greater than according to the undulatory theory. For on the emission theory the velocity of light must be greater in the water telescope, and therefore the angle of aberration should be less; hence, in order to reduce it to its true value, we must make the absurd hypothesis that the motion of the water in the telescope carries the ray of light in the opposite direction!

Airy's experiment was taken as setting back by a long step Newton's corpuscular emission theory of light while confirming Fresnel's ideas of an ether moving inside the moving denser media, though Airy's may have been but another famous negative (null) result arising from misplaced presuppositions.

The angle BCA is formed at the star between the vector CA and CB. In order to change these vectors, it is necessary to change the velocity of light in the entire medium, in the entire space between the star and the observer. Had this value of velocity CA been changed, then indeed the angle would have changed. But changing the velocity value in the last 35.3 inches of light's travel from the star and expecting this to change the angle of aberration that already existed at the distant star—this was not warranted.

REFERENCES

1. Bradley, J.: A new apparent Motion discovered in the Fixed Stars; its cause assigned; the Velocity and equable Motion of Light deduced. *Phil, transact.* 35;637, 1728.

2. Sarton, G.: Discovery of the Aberration of Light. *Isis* 16;233, 1931.

3. Hogben L. Science and the Citizen. 1958; New York, Knopf: 330-336.

4. Doppler C. Ueber die bisheringen Erkläruungsversuche des Aberrations-Phenomens. *Abh.d.Böhmischen Ges.d.Wiss.* V. Folge. vol.3, 1844.

5. Lodge O. Aberration problems. Proceedings Roy. Soc. London, vol. 51; 1892: 98:101

6. Liebscher D. E. Aberration and Relatiity. *Astron.Nachr.*319(5); 1998:309-318.

7. Whyte L. L. Roger Joseph Boscovich. London, G. Allen & Unwin Ltd, 1961: 177.

8. Stachel J. Fresnel's (Dragging) Coefficient as a Challaenge to 19th Century Optics of Moving Bodies. In: Einstein Studies; Bikhauser Boston, 2005; 11:1-13.

9. Janssen M. 19th Century Ether Theory. @ Scholar.Google; Einstein for Everyone. Fall 2001

10. Biot C. Essais sur la Theorie Mathematique de la Lumiere. Paris. Mallet-Bachelier, 1864: 44-46

11. Airy, G. B. Undulatory Theory of Optics. 1877; London, MacMillan, p. vii.

12. Klinkerfues, W.: Fernere Mitteilungen ueber den Einfluss der Bewegung der Lichtquelle auf die Brechbarkeit eines Strahles. *Nachr. Koenigl. Ges. d. Wissen.*, Goettingen, 4;33, 1866.

13. Foucault ML. Méthode générale pour mesurer la vitesse de la lumière dans l'air et les milieux transparants. *Comp. Rendus* 30: 1850: 551-60.

14. Airy, G. B.: On a supposed Alteration in the amount of astronomical Aberration of Light, produced by the passage of the light through a considerable thickness of refracting medium. *Proc. Roy. Soc.*20 ; 35, 1871. Also: *Phil. Mag.* 43;210, 1872.

Speed in Both Directions

MICHELSON'S EXPERIMENT

By determining that the speed of light did not change when both its source and its observer were at steady distance from one another and moving forward uniformly on the same platform, Michelson confirmed that light's motions were no different than those of any other motion.

Take a man inside a square boxcar of length and width D (figure) playing billiards, which Michelson did exceedingly well. In this car, he simultaneously shoots two balls with equal velocities c, referred to the car, and at a right angle to one another, say, one to the front and the other to the side. The balls will each reach the opposite wall at the same time T because D and c are equal: $T = T' = D/c$. They will also return to the man at the same time, presupposing that the velocity of going equals the velocity of return.

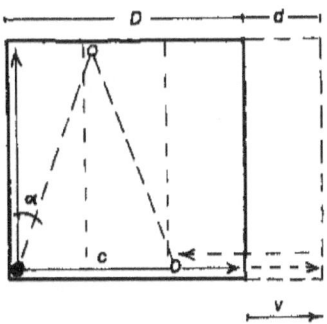

Let the car move forward with uniform velocity v, referred to the road. This uniform motion does not matter to the man inside—when the car is enclosed he cannot even tell whether he is moving at all. When the balls are then thrown, nothing changes as far as the man inside is concerned; the length

D is the same as before the motion, the velocity c referred to the car is the same, and the balls will return to him in the same direction at the same time as when the car was stationary. This all has already been shown in detail by Oresme, Galileo, and Newton.

Let us now see how the event appears to an outsider stationed next to the road. In reference to this man, the velocity of the ball, moving forward inside the moving car, is $c + v$. By the time the ball reaches the opposite wall ($T = D/c+v$), the wall itself has moved a certain distance d. The time of the ball's arrival at this wall is hence $T = (D+d)/(c + v)$.

The ball thrown to the side moves with velocity $c' = v/\sin\alpha$, where $\tan\alpha = v/c$. By the time it arrives, this wall itself moved forward a distance d. The distance this ball travels is $D' = d/\sin\alpha$, and the time of its arrival is $T' = D'/c' = d\sin\alpha/v\sin\alpha = d/v$. This time coincides with the time the other ball took to reach the front wall: $T - T' = 0$; $T = T'$; $d/v = (D + d)/(c + v)$; $d(c + v) = v(D + d)$; $D/c = d/v$, which is true by definition (d is the distance the car covers at velocity v in the time the ball covers distance D at velocity c).

The returning ball from the front wall has the velocity $c - v$ (because the wall recedes at v) and covers a distance $D - d$. Its time of arrival at the man is $D - d/c - v$. The returning ball from the side travels with velocity $c' = v/\sin\alpha$ and covers the distance $D' = d/\sin\alpha$. Its time of arrival is d/v, and this equals the time of arrival from the front wall: $d/v = (D - d)/(c - v)$;

$d(c - v) = v(D - d)$; $D/c = d/v$. The time of return equals the time of going toward the walls. There is no difference between the arrival times of the balls, either going or returning, in direction of the motion or perpendicular to it, either in reference to the man inside the moving car or in reference to the stationary outsider.

In reference to the outsider, the ball that went and returned from the side covers a longer distance (= $2D/\cos\alpha$) than the ball that went forward and back (= $2D$). The balls nonetheless meet again at the same time because the velocity of the ball that traveled sideways, referred to the road ($c/\cos\alpha$), is higher than the velocity of the ball that went and returned in the line of motion of the car (= c). (See also analysis of the interpretations to Michelson's experiment at the end of this chapter.)

Albert A. Michelson (1852-1931) accepted Maxwell's belief that the speed of light in direction of the earth's movement was not equal to its speed in the opposite direction and that this difference could be detected on the earth itself. He thought with Maxwell that the success of these efforts to determine the earth's motion by experiments on its surface failed due to inadequately sensitive measuring devices and trusted that with the help of his newly invented interferometer he could do it.

"On the hypothesis of a stationary ether it appeared possible to detect a motion of the earth independent of astronomical observations." He trusted this fundamental possibility, the basis of the whole project, at the beginning of his 1881 paper: "Assuming then that the ether is at rest, the earth moving through it, the time required for light to pass from one point to another on the earth's surface, would depend on the direction in which it travels." (figure)[1]

Michelson's experiment was often heralded as "one of the most famous experiments in Physics"[2] or as "the greatest of all negative results in the history of science."[3] Having attained this inordinate fame, we may not be judged completely amiss when casting a fresh look at it, projected here on the background of our knowledge about motion in and between inertial frames of reference as detailed in the previous chapters. The entire first page of his 1881 article is here reproduced for easy reference.

ART. XXL—*The relative motion of the Earth and the Luminiferous ether;* by ALBERT A. MICHELSON. Master, U.S. Navy.

THE undulatory theory of light assumes the existence of a medium called the ether, whose vibrations produce the phenomena of heat and light, and which is supposed to fill all space. According to Fresnel, the ether, which is enclosed in optical media, partakes of the motion of these media, to an extent depending on their indices of refraction. For air, this motion would be but a small fraction of—that of the air itself and will be neglected.

Assuming then that the ether is at rest, the earth moving through it, the time required for light to pass from one point to another on the earth's surface, would depend on the direction in which it travels.

Let V be the velocity of light.

v = the speed of the earth with respect to the ether.

D = the distance between the two points.

d = the distance through which the earth moves, while light travels from one point to the other.

d_1 = the distance earth moves, while light passes in the opposite direction.

Suppose the direction of the line joining the two points to coincide with the direction of earth's motion, and let T = time required for light to pass from the one point to the other, and T_1 = time required for it to pass in the opposite direction. Further, let T_0 = time required to perform the journey if the earth were at rest.

$$\text{Then } T = \frac{D+d}{V} = \frac{d}{v} \; ; T_1 = \frac{D-d}{V} = \frac{d_1}{v}$$

From these relations we find $d = D\dfrac{v}{V-v}$ and $d_1 = D\dfrac{v}{V+v}$ whence T

$= \dfrac{D}{V-n}$ and $T_1 = \dfrac{D}{V+v}$; $T - T_1 = 2T, \dfrac{v}{V}$ nearly, and $v = V\dfrac{T-T_1}{2T_0}$.

If now it were possible to measure $T - T_I$ since Y and T_o are known, we could find v the velocity of the earth's motion through the ether.

In a letter, published in "Nature" shortly after his death, Clerk Maxwell pointed out that $T - T_I$ could be calculated by-measuring the velocity of light by means of the eclipses of Jupiter's satellites at periods when that planet lay in different directions from earth; but that for this purpose the observations of these eclipses must greatly exceed in accuracy those.

Michelson had a very sensitive instrument constructed by Schmidt and Haensch in Berlin, where he resided at the time (figure), consisting of two equally long arms with mirrors (c, b) at one end of each. Light from the flame of a small lantern (s) was split in two by a plane-parallel plate (a); and the recombined beams, reflected from the mirrors, were observed at e. When returned, the beams refracted by the plate and reflected by the mirrors formed fringes, bands of brighter and darker light. If the speeds of going forward with the earth's direction were different when the whole apparatus was rotated perpendicular to the earth's motion, then these fringes would shift, be in different position. Light was first transmitted when arm ac was in direction of the earth's motion while arm de was perpendicular to it, and then the instrument was rotated at a right angle and the measurement repeated.

The Institute of Physics in Berlin proved too shaky for the sensitive instrument

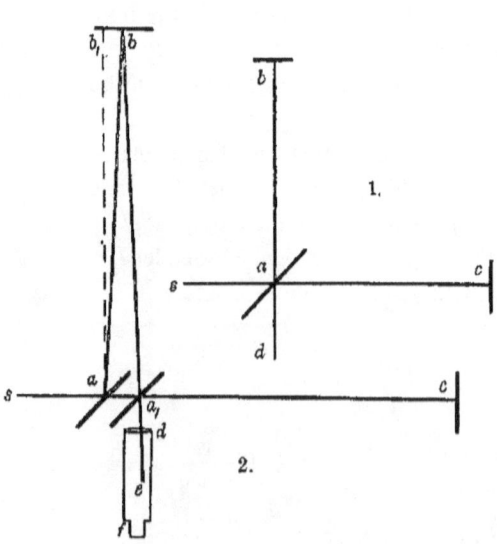

due to vibrations from the local streetcars, so Michelson moved to the suburb of Potsdam where, in a cellar of the Astrophysical Observatory, late into the quiet night, he carried out his observations. These were repeated many times, later with the help of Edward V. Morley in Cleveland.

To Michelson and everyone else's amazement, puzzlement, and disappointment, the fringes did not shift; that is, there was no difference at all in the times light took to travel in direction of the earth's motion and in direction perpendicular to it.

We start our analysis with Michelson's definitions and arithmetic. On the first page of his first paper (see above), he said[1] "Let V be the velocity of light." The velocity of light in reference to what? As Michelson himself often measured, it was always understood to mean the speed between two *stationary* positions on earth or, as Roemer did, in reference to a fairly stationary earth in reference to Jupiter such as the velocity of light measured when the earth was in fairly steady position closest to Jupiter and when farthest away from it, about 300,000 km/sec (In general, and by definition, Velocity = Distance/Time, and $T = D/V$).

Said Michelson, "D = the distance between two points. d = the distance through which earth moves, while light travels from one point to the other." And here was the crux of the problem.

D is the distance between two points on earth. The d is the distance the earth moves in reference to what? As previously noted, it is of paramount importance to continually keep in mind that the motion of the earth in this and similar instances can only be considered from a stationary position outside the earth, such as the "man in the moon" according to Kepler,[4] or "man in heaven" in Oresme's words. Michelson's saying "Assuming then that the ether is at rest, the earth moving through it" is tantamount to the earth moving in reference to some other generally accepted stationary object or medium outside it, such as the stars or the sun. In what other point of reference was the ether at rest? The decisive difficulty was that Michelson was not observing the phenomena from a stationary position outside the earth or outside the ether. In the ether or on the earth itself, without an external point of reference, the distance "d" he was talking about cannot be determined.

The v = the speed of the earth with respect to the ether [at rest]." Now, therefore, if the earth moved in reference to an external stationary medium or object, the velocity of light traveling in the same direction on this earth, as employed by Michelson in his experiment and as seen from the stationary position in outer space, would necessarily be V + v, not simply v. No evidence whatsoever existed in Michelson's time to permit neglect of compounding the velocity of light with the velocity of the source or observer; on the contrary, Roemer and Bradley have already furnished the necessary data in support of the need to do it.

Furthermore, if "T = the time required for light to pass from one point to the other" (on the moving earth, equals D/V) and light's velocity in reference to a stationary position outside earth increased by the earth's velocity, then it should be $T = (D + d)/ (V + v)$. But instead, Michelson wrote, "$T = (D + d)/ V$." The velocity of light (V) was a priori not compounded by the velocity (v) of the source on the earth moving in reference to the stationary ether or sun while the distance covered ($D + d$) was indeed reckoned in this frame. *The events were considered confusedly from two different points of reference and, therefore, could not possibly correspond.*

At this juncture, the reader may perhaps object, for even though Michelson spoke of the *time* required for lights to pass in different directions, he did not really stand there with a stopwatch to measure times. What his interferometer was supposed to measure was differences in *lengths* in terms of wavelengths so that, even granted that, the times of going and return in both directions were the same; the distances certainly were not. In the figure, the solid line represents a wave of length AB while the interrupted line CD represents a wave of the 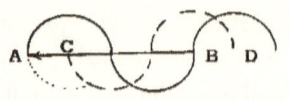 same length, but, say, half a distance behind and in opposite phase. When point C coincides with point A, the waves "interfere" and cancel one another, according to Thomas Young, MD, so that a dark fringe is visible in the spectrum. When the waves do not exactly coincide, the fringe is displaced, and this is what Michelson tried to detect: Did the fringes change when light traveled different distances on account of the earth's motion in reference to the stationary sun or the ether? If so, the rate of the earth's motion could be deduced.

At the bottom of the second page of his first paper of 1881, he explained, "The pencil which has traveled in the direction of the earth's motion will in reality travel 4/100 of a wave-length farther than it would have done were the earth at rest. The other pencil being at right angles to the motion would not be affected." However, as noted above, the pencil will only travel farther forward when seen from a point outside the earth in reference to which it is at rest or moving, not from a position on it, where Michelson and the reader of his report were in fact positioned.

The paper ended with the following conclusion: "The result of the hypothesis of stationary ether is thus shown to be incorrect, and the necessary conclusion follows that the hypothesis is erroneous. This conclusion directly contradicts the explanation of the phenomenon of aberration which has been hitherto generally accepted, and which presupposes that the earth moves through the ether, the latter remaining at rest."

The next paper on the subject written with Edward Morley begins also with this problem of aberration:[5]

> The discovery of the aberration of light was soon followed by an explanation according to the emission theory [Bradley]. The effect was attributed to a simple composition of the velocity of light with the velocity of the earth in its orbit. The difficulties in this apparently sufficient explanation were overlooked until after an explanation on the undulatory theory of light was proposed. This new explanation was at first almost as simple as the former. But it failed to account for the fact proved by experiment that the aberration was unchanged when observations were made with a telescope filled with water [Airy]. For if the tangent of the angle of aberration is the ratio of the velocity of the earth to the velocity of light, then, since the latter velocity in water is three-fourth its velocity in a vacuum, the aberration observed with a water telescope should be four-thirds of its true value.

We now return to see how Michelson dealt with this light at right angles to the direction of the earth's motion. In 1881, he said, "If, however, the light had traveled in a direction at right angles to the earth's motion it would be entirely unaffected, and the time of going and returning would be, therefore, 2 $(D/V) = 2\ T_0$." ($T_0 =$ time passed when the earth is assumed to be at rest.) He did not explain why it was entirely unaffected.

In the winter of 1881, Michelson went to Paris to study at the Collège de France and the École Polytechnique and to demonstrate his apparatus to some famous physicists in Paris, such as Cornu, Mascart, and Lippmann. He also demonstrated his experiment to the Paris Académie des Sciences at the February 20, 1882, meeting, subsequently published in French; and there he admitted to an error: "Dans ce Memoire [paper of 1881] j'ai oublié l'efet du mouvement sur le rayon *bc* [the perpendicular]. La correction m'a été signalé par M. Potier."[6] More clearly expressed in the 1887 paper,

> In deducing the formula for the quantity to be measured, the effect of the motion of the earth through the ether on the path of the ray at right angles to this motion was overlooked . . . It may be mentioned here that the error was pointed out to the author of the former paper by M. A. Potier, of Paris, in the winter of 1881.

To my knowledge, nothing has been published about Potier since his obituaries, and nothing at all in English. A short biographical note may, therefore, not be completely out of place, particularly since he played a pivotal role in the interpretation of Michelson's experiment. Alfred Potier (Paris, May 11, 1840-May 8, 1905)—Swenson named him André[7]—was a French polymath who contributed to many theoretical and practical fields of science when this was rapidly expanding. His interests covered mainly mathematical physics, the nature of light and the ether, geology, and electricity and magnetism and their practical applications in industry. Born in 1840, Potier entered the legendary École Polytechnique at age seventeen where, in 1867, he became physics teacher and then, in 1881, full Professor of Physics, succeeding Jamin and preceding Nobel laureate Henri Becquerel.[8] At the same time, he was member of the State Mining Engineers Corps, occupying the chair of physics in the École des Mines where he taught Henri Poincaré. Geological works included revisions of the geological map of France and submarine topographies in Pas-de-Calais in order to examine the feasibility of a tunnel to England. These, and his valor during the German siege of Paris in 1870, earned him the Legion of Honor. His other publications concerned Fresnel's theories of light and the ether,[9] diffraction of polarized light, elliptical reflection, magnetic rotational forces, or interference fringes. He helped translate and contributed extensive notes to J. C. Maxwell's treatise on electromagnetism,[10] facilitating its reading in France. The French Physics Society appointed Potier as president in 1884; the International Electricians Society did the same in 1895. In 1891, he was accepted into the French Academy of Science.

Potier was a member of many committees at the famous 1881 Universal Exposition in Paris, including the one that set the standards for units in electricity such as the ampere, volt, and ohm. It was in 1881 in Paris that he met Michelson and expressed his view that the latter's calculation in his aether drift experiment was in error, and that if corrected, the discrepancy in the times elapsed in the two routes would be completely eliminated. In a letter to Lord Rayleigh on March 6, 1887, Michelson wrote that he "had an indistinct recollection" of Lorentz mentioning the same correction. "I have not yet seen Lorentz' paper and fear I could hardly make it out when it does appear."[11]

In the 1887 paper, written with Edward Morley, Michelson accordingly amended the route of the perpendicular light from ab_1 to ab (figure 2). When the instrument on the moving earth is observed from a stationary point outside earth, the mirror at b_1 already moved to b while light was traveling there from a. "This meant that Michelson had overestimated by a factor of two the fringe shifts originally expected."[7] The mirror at c also moved forward at the same time, but this curiously was not taken into account. It thus seemed that light traveled different distances at the same time.

The obscure M. Potier in Paris clearly viewed the experiment on earth from a stationary position outside; if one considered with Michelson "$d =$ the distance through which the earth moves, while light travels from one point to the other," carrying with it the perpendicular mirror, one ought also consider the distance b_1b (figure 2, left) through which light traveled while in the perpendicular path to this displaced mirror. Compared to the diagram of 1881 (and figure 1, right), the perpendicular ray in 1887 did not go perpendicularly to b_1 but a bit forward to position b (figure 2). "The angle bab_1 being equal to the aberration $= \alpha$. Let it now be required to find the difference in the two paths aba_1, and aca_1."

The distances covered are unequal, and when one neglects to compound the velocities in the two directions perpendicular to one another, the times of return are indeed unequal. Given, however, that the compounded velocities are in fact unequal, the times of going forward and return and going sideways and return are certainly equal, and no shift in interference fringes should be expected. The speed forward is $V + v$ and return in $V - v$. The speed perpendicularly is the same going and return. It is $2(v \tan \alpha)$, or $2\sqrt{V^2 + v^2}$, which is larger than simply V and equals $(V + v) + (V - v)$. (See figure below under Interpretations.)

The angle of aberration in Bradley's case was formed by a moving observer on earth in reference to the stationary source—the star. When Michelson's case is viewed from a stationary point outside earth, the angle was formed by the moving light source in reference to this stationary observer, which, as we know since Oresme and Copernicus, is the same thing. The compounding of the velocity of light by the velocity of the observer created the angle of aberration in both observations, and the value ab in Michelson's case was certainly larger than ab_1, just as Bradley's velocity CB was higher than CA. Whatever moved the light from position b_l to b in the experiment was imparted to it by the motion of the earth, the same motion that moved the observer Bradley from position B to A. Had the light from the source not been compounded by the earth's motion (momentum), it would have gone perpendicularly to b_l and, thus, missed the mirror, which was already a little forward.

Michelson then accepted the fact that the motion of the earth altered (increased) the motion of light in the perpendicular direction: "In consequence, the quantity to be measured had in fact but one-half the value supposed. If, as was the case in the first experiment, $D = 2 \times 10^6$ waves of yellow light, the displacement to be expected would be 0.04 of the distance between interference fringes."

The wavelength difference was diminished by half, and yet, the two rays were still a bit out of phase; and theoretically, the fringes should have shifted, but in reality they did not, the expected displacement did not occur!

When, in 1887, it was admitted in this manner that the earth's motion influenced the distance and speed of light in its *perpendicular* direction, it may be seen as no small oversight not to have gone back and corrected the 1881 calculations for the *forward* direction as well. And yet the definitions have not changed: "Let V = velocity of light," that is, the light emanating from the source on the moving earth. Now if the velocity of the transverse ray was compounded by the earth's motion, it must do so also in direction of its forward motion, and velocity V was in reality $V + v$ (as pointed out previously). When incorporating this correction into the calculations, the other "one-half of the value supposed" is found, and the two opposing rays portrayed do indeed cancel one another, and the null shift in fringes comes as no surprise or disappointment.

Michelson concluded,

> It appears from all that precedes, reasonably certain that if there be any relative motion between the earth and the luminiferous ether, it must be small; quite small enough entirely to refute Fresnel's explanation of aberration . . . It is obvious from what has gone before that it would be hopeless to attempt to solve the question of the motion of the solar system by observations of optical phenomena *at the surface of the earth* [his emphasis].

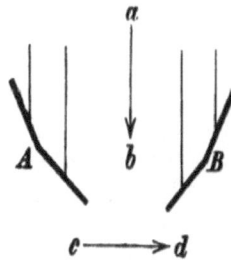

Not only "by observations of optical phenomena,"—for in a uniformly moving ship or the earth as stressed by Oresme, Galileo or Newton—in an "inertial frame of reference," if you wish, *all* uniformly linear motions are the same. The first method of detecting any motion of the earth, its daily *angular spin*, by observing motions on its own surface was discovered in 1851 by Jean Foucault with his pendulum.

Ernst Mach, in 1873, already suggested a method of determining the motion of the earth by using the sun as the source of light rather than a source on the moving earth. The sunlight arrives from direction *ab*. The earth moves in direction *cd*. Light reflected from mirror A interferes with that from mirror B to form fringes. These fringes (or absorption lines) would shift with change in the direction of the earth's motion.[12]

Interpretations

Reasoning by analogy compounded the confusion in subsequent interpretations of Michelson's experiment. As previously done by equating light to sound, analogies are always fraught with danger, for the comparison is hardly ever precise and, sooner or later, falls apart little by little. Here, the motions of the two light rays were compared to the motions of two swimmers or boats in a moving stream. Quoted by his daughter, Michelson himself made the analogy:

> Two beams of light race against each other, like two swimmers, one struggling upstream and back, while the other covering the same distance, just crosses the river and returns. The second swimmer will always win, *if there is any current in the river.*[11]

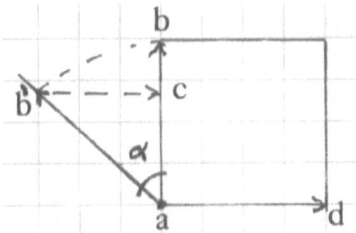

As a naval officer and a competent small-boat sailor, the analogy came to Michelson naturally: Say the river's current (b to c) is 3 meters/sec and the swimmers move at 4 m/sec. The horizontal one goes from a to d at speed $4 + 3$ m/sec; he gets there in 4/7 sec = 0.57; on return from d to a, his speed is 4 - 3 and reaches a in 4 seconds, totally 4.57 sec. In order for the vertical swimmer to go to b, he must swim in direction b' and, in one second, reaches point c on line ab, 2.65 meters from a. His speed in direction to b is 2.65 m/sec, and his total time to b is about 1.5 sec. His total going and return trip thus takes about 3 seconds, faster than going horizontally (4.57). (Approximately, sin α = 0.75; α = 48.6°; ac = 4 cos α = 2.65 or $\sqrt{4^2 - 3^2} = 2.65$.) The swimmer going across returned faster than the one going down and upstream, yet in Michelson's experiment, the rays returned at the same time!

Other authors readily used Michelson's analogy,[13] so for instance, Sir Bertrand Russell, famous philosopher, mathematician, and author with Whitehead of the well respected *Principia mathematica*:[14] "Now anybody can verify, either by trial or by a little arithmetic, that it takes longer to row a given distance on a river up-stream and then back again, than it takes to row the same distance across the stream and back again."

The same oarsmen on the river were similarly described by Clement Durell, forwarded by Freeman Dyson, adding, as Russell suggested, "a little arithmetic" and a diagram (figure).[15]

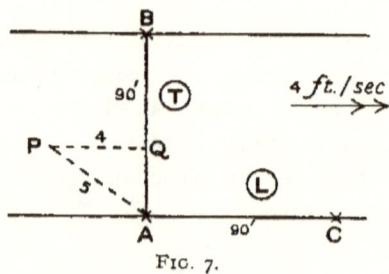

FIG. 7.

A stream is flowing at 4 feet per second between straight parallel banks 90 feet apart. Two men start from point A on one bank; one of them T rows straight across the stream to the opposite bank at B and returns to A, the other L rows to a point C 90 feet downstream and then rows back to A. Each of them rows at 5 feet per second relatively to the water. Compare their times.

Durell concluded, "It therefore takes longer to go down and up than equal distance across and back." Bernard Jaffe,[16] in the same vein, talked about a swimmer going across and up and down the river, an analogy also attributed to Einstein (figure).[17]

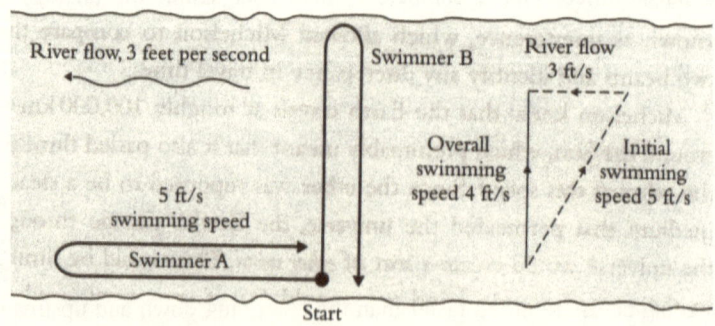

Herman Bondi,[18] in the figure below (note the direction of the ship), used a ship as example of the moving ray.

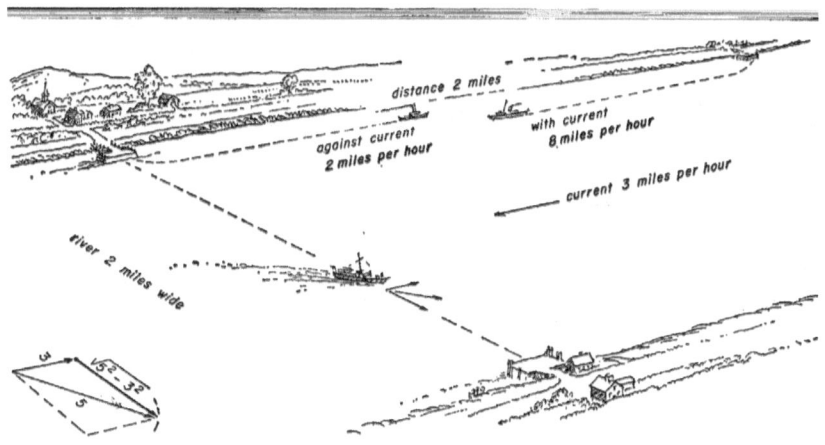

The first discordance of the analogies with the actual experiment arises from the fact that, contrary to the swimmer or boat going across, the ray in the experiment was not directed upstream but straight perpendicularly. The ray does not know that it wants to go across and, for this purpose, must shoot upstream. It is simply shot perpendicularly. Secondly, and decisively, Michelson's laboratory with its interferometer, the light source, and the final observer peering into the telescope all moved uniformly with the earth and were all observed from this moving earth, not from some extraterrestrial fixed stationary position in space above. The rowers, swimmers, or ships, on the other hand, all started from a fixed position on the stationary river bank; and they moved and were observed in reference to this fixed position, say, from shore or a bridge over the river.

Therefore, a more accurate rendition of the experiment, illustrated earlier by the moving balls, is offered numerically in the following figure: On a large ship, securely tied to the dock, is a pool ab_1CD (solid lines) measuring, for the sake of simplicity, four square meters. A fish swims from position a to b_1 at a speed of 4 m/sec while, at the same time, another swims toward position C also at 4 m/sec. Both traveling with equal

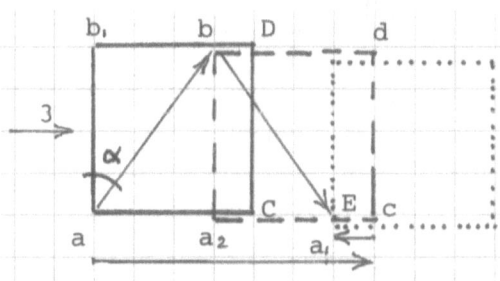

speed will, of course, reach their destinations in one second and will return to position a at the same time, both as measured from a position on the stationary ship and from a bridge stationed over the river.

Now let the ship cast off and drift downriver in direction

of C at a rate of 3 m/sec from a position marked in the figure on the left by solid lines to one marked by interrupted ones. To the person stationed on the ship, the swimmers still behave in the same manner; they move the same distances (total of 4 + 4 = 8) in either direction in the pool and return at the same time. However, to an observer stationed fixed on a bridge over the river, the motions and distances are different; they are compounded by the motion of the ship. In one second, the swimmer going across toward b_1 arrives at position b, 3 meters downstream from position b_1, but still exactly opposite the starting point that is now at position a_2. The *distance ab* (5 meters) is longer than ab_1 (4 meters), ($ab = \sqrt{ab_1^2 + b_1b^2}$), (= bb_1 sin α), but the *speed ab* (taken as a vector magnitude) is also higher, being a compound (increase) of the speed in direction ab_1 with the speed in direction b_1b; and the swimmer will, therefore, arrive at b in one second of time, the same as when the ship stood still.

Downstream, the swimmer goes at the speed 4 m/sec, compounded (increased) in the same direction by the speed of the water, 3 m/sec, that is, 7 m/sec. In one second, he will arrive at c, 7 meters away, Thus, in one second, the cross swimmer covered a distance of 5 meters, and the downstream one, 7 meters; but they arrived at the same time, to the same corners of the pool (b and c) because their speeds were correspondingly different. The angle α is the angle of aberration mentioned by Bradley and Michelson.

On return from across position b, the swimmer arrives at a_1 in one second, the same as from a to b (total distance aa_1 is 10 meters). The other swimmer, however, returning upstream against the current, travels at only 1 m/sec (4 - 3) and arrives at a_1 (total distance of 7 + 1 = 8)—also in one second—the same time as the first swimmer. In two seconds, both traveled from a to a_1 (six meters apart)—one across the stream, the other, up and down the stream; that is, different distances (10 vs. 8) but both arrived to the corner of their departure (a and E) at the same time! The results do not change when the swimmer first goes upstream instead of down except that, then, his speed is reversed; he swims slower up and faster down. The cross swimmer's speed remains the same.

When the proper frame of reference is taken into account, there is no difference between the arrival times, no shift on the clock's arms. Either going or returning, in direction of the motion or perpendicular to it, either in reference to an observer on the moving ship or in reference to the stationary outsider. These are facts that are repeatable, consistent, and fairly easily verified—generally applicable kinematic principles. As applied to real bodies, their mass must be considered. Applied to light, its altered speed by reflection may also be taken into account as outlined in the optokinetic part. The dimensions in Michelson's experiment compared to astronomical ones are, however, so miniscule that these limitations probably carry little weight at this stage of our knowledge.

For the purpose of detecting the relation between the speed of light and the speed of its source or its observer, three methods were already known in Michelson's time—those of Roemer, Bradley, and Doppler. They were confirmed by Michelson's experiment; namely, the motions of light were no different than any other linear motions and must be compounded by the motion of the observer or source. All data available up to the year 1900 supported this conclusion. Michelson investigated the ether and found that it was not stationary in reference to the earth and not stationary in reference to the sun. There was no way to elucidate the nature of this ether, and the talk about it was thereafter transplanted from physics books to those on history.[19, 20, 21] His results seemed to support Newton's concept of material light corpuscles, an anathema in the nineteenth century.

Yet with the death of the ether, the wave was not buried but was newly reborn with an electromagnetic soul clothed in mathematical garments. The ether may have departed from the physical realm, but to mathematicians (who by nature of their profession are less closely attached to experiments and real experiences), the wavy motion had the immutable force of dogma. "The conclusive null result of the Cleveland experiment was decisive in its influence on Lorentz, FitzGerald, Larmor, Poincaré, and Einstein in developing their theories of the electrodynamics of moving bodies, which culminated in the special theory of relativity."[22]

Joseph Larmor (1857-1942), fellow Royal Society, winner of many prizes in physics and mathematics, reconstructed the experiment in the next figure.[23] The whole arrangement now is seen and calculated from a stationary point outside the page. The speed of light he marked with V and that of the mirrors (A and B) with v. Because mirror A was receding, he said the speed of light "from G to A is V - v and from A back to H is V+v." The velocity along the lines GB = BH = V. This is a bit confusing. If V is the original speed of light, then its speed along GB must be compounded (somewhat increased) by v. In addition, mirror B is also receding, yet this was not taken into account. But if the speed of light changed because the mirrors were receding, source S was, at the same time, advancing; hence, the speed of light from this source, as seen from a stationary point outside the earth, ought to have been set at V+v, not simply V. One cannot have it both ways.

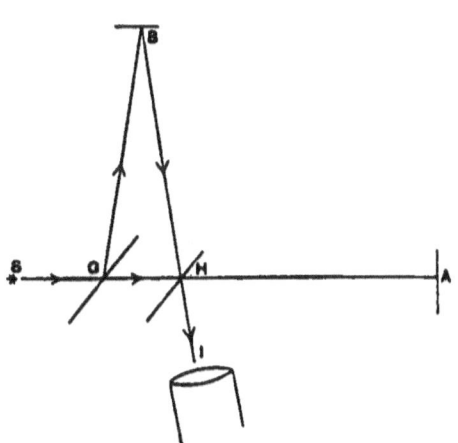

Albert Einstein (1879-1955) also became interested in Michelson's experiment in as much as it was frequently discussed by his colleagues, most influential among them being Hendrik Lorentz. The latter is generally regarded as the first to pay serious attention to the theoretical consequences of Michelson's experiment and to have largely derived his theories from it.[24] His many papers were widely quoted, and his book-length article (280 pages) in the *Encyklopädie der mathematischen wissenschaften* of 1903 was probably required reading by serious mathematicians such as Einstein, given that Lorentz received the Nobel Prize the previous year. In this treatise too, Lorentz discussed Michelson's experiment (page 173).[25]

Being a pure theorist, Einstein himself paid but cursory attention to experiments in general and to this one in particular. No need for observations, "physics is an attempt conceptually to grasp reality as it is thought independently of it being observed." And "a theory can be tested by experience, but there is no way from experience to the setting up of a theory."[26]

Nonetheless, Einstein could not escape dealing with Michelson, as Nobel laureate (1918) Max Planck put it[27] (p. 60), "The Theory of Relativity was led up to by Michelson's experiments on optical interference." In his 1915 article "Die Relativitätstheorie,"[28] Einstein pictured Michelson's experiment (figure), calling it "ein wichtiges experimentelles resultat"—an important experimental result—and then marches forth to explain how his theory developed from it through the FitzGerald and Lorentz contraction hypothesis.

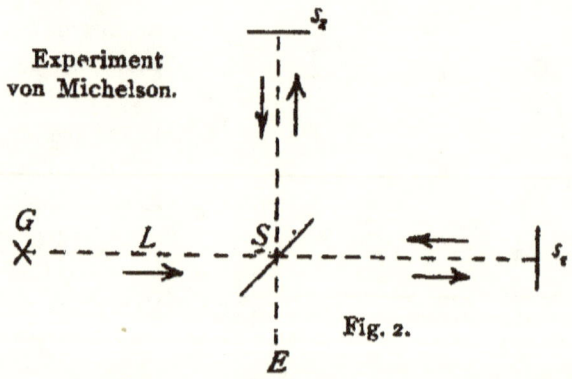

Fig. 2.

The lengthy article Einstein wrote in 1907 (published the following year) "On the Relativity Principle and its Consequences"[29] also begins with a discussion of Michelson's experiment; and from there, reasoning through a large number of mathematical equations, he concluded with the famous equation $M = E/c^2$. This equation, he said, applied not only to inertia but also to gravity of mass (I don't know who transposed it to the popular $E = mc^2$). The paper may have been composed in some haste, for its author was obliged to publish a follow-up with no less than seven corrections.[30]

Albert Einstein and Albert Michelson seem to have been, to use Henry Longfellow's words, "ships that pass in the night, and speak each other in passing" (figure, note direction of gaze).

For whereas Einstein was solving nature's problems strictly with pencil and paper, Michelson was fussing with optical apparatus and precise measurements. He openly admitted and regretted his weakness in mathematics. Einstein's Nobel Prize of 1921  was awarded "for his services to Theoretical Physics, and especially for his discovery of the law of the photoelectric effect." Thanks to Nobel laureate (1911) ophthalmologist Allvar Gullstrand, the theory of relativity was not judged fit for inclusion in the prize. Michelson's award of 1907 was "for his optical precision instruments and the spectroscopic and metrological investigations carried out with their aid." His reputation for precision was so overwhelming that his faulty calculations were not easily questioned, particularly after the Nobel.

REFERENCES

1. Michelson Albert Abraham, "The relative motion of the Earth and the luminiferous ether," *American Journal of Science,* series 3, xxii (1881), 120-29.
2. Bondi, H.: Relativity and Common Sense. New York, Doubleday, 1964, p. 54.
3. Jaffe, B.: Michelson and the Speed of Light. New York, Doubleday, 1960, p. 88.
4. Rossmann Fritz, *Nikolaus Kopernikus, Erster Entwurf seines Weltsystems* (Hermann Rinn, Munich, 1948), p. 12.
5. Michelson Albert A., Edward W. Morley, "On the relative motion of the earth and the luminiferous ether," *American Journal of Science*, 3rd series, xxxiv (1887), 333-45
6. Michelson A. Sur le movement relative de la Terre et de l'ether. *Comptes Rendus*; Paris; 20 February 1882: 520-23.
7. Swenson LS. *The Ethereal Aether.* Austin. University of Texas Press, 1972; p 73.
8. Lienard A. 1908. "Alfred Potier." http://www.annales.org/archives/x/potier.html. Retrieved 9/9/09.
9. Potier A. "Consequences de la formule de Fresnel à l'entrainement de l'ether par les milleux transparantes." *Journal de Physique* (Paris); 3: 201-4. 1874.
10. Maxwell JC. 1885. *Traité d'electricité et de magnétisme.* A. Curnu, A. Potier, E. Sarrau (eds.). Paris; Gauthier-Villars.
11. Livingston, DM. The Master of Light. New York, Charles Scribner's Sons; 1973: 124.

12. Mach E. Beiträge zur Doppler'schen Theorie der Ton-und Farbenanderung durch Bewegung. J. G. Cale; Prag,1873.

13. The following discussion was largely presented as a poster titled 'The Common View of Michelson's Experiment' at the November 2008 annual meeting of the History of Science Society in Pittsburgh, PA.

14. Russell Bertrand, *The ABC of relativity,* (New York, 1958), 26-27.

15. Durell Clement V., *Readable relativity*, (New York, 1966), 25-27.

16. Jaffe Bernard, *Michelson and the speed of light*, (New York, 1960), 64-66.

17. Singh S. *Big Bang*, The Origin of the Universe. New York, Fourth Estate; 2004.

18. Bondi Hermann, *Relativity and common sense*, (New York, 1964), 55-57

19. La Rosa, M.: Der Aether, Geschichte einer Hypothese. Leip zig, Barth, 1912.

20. Schaffner K. F.: Nineteenth Century Aether Theories, New York,1972; Pergamon Press.

21. Cantor G. N., Hodge M. J. S.: Conceptions of ether. Cambridge, 1981; Univ. Press.

22. Shankland R. S. Michelson-Morley experiment. *Am. J. Physics,* 1964: *xxxii:* 16-35.

23. Larmor J. Aether and Matter. Cambridge, University Press 1900: 46-53.

24. Lorentz H. A. Electromagnetic Phenomena. In: the Principle of Relativity. New York, Dover Publications 1923: 9-34

25. Encyklopadie der Mathematischen Wissenschaften. A. Sommerfeld ed. Leipzig; B. G. Teubner; 1903, vol 5 (2):67-290.

26. Schlipp P. A. ed. Albert Einstein: Philosopher-Scientist. New York, Harper & row, 1959: vol. 1: 81, 89.

27. Planck M. The Universe in the Light of Modern Physics. New York, WW Norton, 1941:60.

28 Einstein A. Die Relativitatstheorie. In: Die Kultur der Gegenwart. Leipzig & Berlin. B. G. Teubner, vol 3 (1);1915:703-713.

29. Einstein A. Über das Relativitatsprinzip und die aus demselben gezogenen Folgerungen. *Jahrbuch d. Radioaktivität u. Elektronik.* vol 4, 1908: 411-462.

30. Einstein A. Berichtingungen zu der Arbeit "Über das Relativitätsprinzip," *op cit* vol. 5; 1908: 98-99.

DEFINITIONS

Definitions

Names and words often mean different things to different people, and since we cannot smell them, a rose may not be a rose by any other name. Definitions are essential in telling us who is who and what is what so that we do not confuse the elements in the story. Our story was an attempt on our part to understand an event in nature—light—and we finish by introducing the terms by which we defined nature. "The first cause of absurd conclusions I ascribe to the want of method; in that they begin not their ratiocination from definitions."[1]

This quest to define the natural world—to tell the nature of Nature—is as old as civilization itself and was usually in the purview of *Metaphysics* (originally a book written by Aristotle after his *Physics*) that treats mainly of topics unrelated to experience, best known among them being mathematics.[2] But for our purposes, we need definitions that relate to the reality of the world as perceived by our senses and conceived by physiological thought processes—defined terms that aid comprehension of real natural phenomena. Thus the definitions here are not presented with a purpose of forming a logical system of axiomatic premises whence our knowledge is then deduced by a strict mental discipline, but are meant merely to describe the milieu in which the events in this volume occur. We attempt to employ Francis Bacon's method of evidence-based epistemology as the proper induction to reliable knowledge, appreciating that this empiricism was derived from the ancient Greek word for practitioners of medicine called empirics.

1. Hobbes Th. Leviathan; Chapter V. 'Of Reason and Science'. Univ. of Chicago, 1952: 59.
2. Planck M. The Universe in the Light of Modern Physics. New York, WW Norton, 1931:113.

Man

The existence of man is axiomatic; namely, it cannot be proved or disproved by man and, hence, must be accepted for granted, a priori. The axiomatic nature of man seems at first puzzling; historians and anthropologists assure us that there was a time when man did not exist, and present developments are sufficient reason to fear that he will soon cease to exist. But these considerations emanate from man himself; it is he who says so. Logically, no thing can possibly be proven or demonstrated by the thing itself, which leaves no choice but to accept the axiom of man.

A lone person on an island cannot prove his existence. Unless some signal from him or his remains is received by the rest of mankind, he does not exist. His reality is not fact nor truth, for there is no way of demonstrating it. He can still prove his existence to the fish, but then the fish must take their own being as an a priori fact. What the man on the island lacks is another human frame of reference. Man may imagine a world without man, but it is man who does the imagining. All human endeavors—including science—begin with man even though man himself cannot be proven or demonstrated.[1]

Early in his history, man evolved a concept named truth, God given and, at least, on one occasion, chiseled in stone on Mount Sinai. When, in the Renaissance, blind emotional faiths were gradually replaced by more enlightened rational thoughts, the concept of transcendental truth remained but its contents changed. Scientists, as theologians before them, often believed that their truths were self-propelled by some innate power of passive buoyancy that, like oil in water, must sooner or later rise to the surface—*vincit omnia veritas*. Belief in this abstract entity was termed by Jacques Monod,[2] "the postulate of objectivity," where objective meant without man as distinct from the subjectively human. It was perhaps best expressed by Ernst Mach: [3]

> If the historical sciences have inaugurated wide extensions of view by presenting to us the thoughts of new and strange people, the physical sciences in a certain sense do this in a still greater degree. In making man disappear in the All, in annihilating him, so to speak, they force him to take an unprejudiced position without himself,

and to form his judgments by a different standard from that of the petty human.

Holding strong convictions, one naturally liked seeing them transcend petty humanity—beyond human doubt and frailty. When, last century, Mach's countrymen took the annihilating a bit more literally, it impressed with horrifying impact the perils of dehumanizing science. The idea of absolute and objective truth generated also absolute and objective laws, which abound in some branches of knowledge and serve perhaps to remind us of Tolstoy's saying that where there is law, there is injustice. The brunt of the endeavor and the aim of science was the discovery of facts, axioms, and phenomena that had existence independent of man while, at the same breath, conceding that it was man doing all this.

It is now generally recognized that human psychological and social factors influence man's perception of reality.[4-9] In order, therefore, to fathom the validity of assertions concerning "true facts," one must allow for the psychological state of those asserting them and the social context—of the society of scientists and society at large[10]—in which the assertions were made. The innocent belief in true reality independent of man has been recently amended to include man; namely, a society of experts. A true physical fact is now often understood to mean a state of affairs that appears in only one particular way to the largest number of interested observers, a process named by Michael Polanyi "mutual authority,"[11] and by John Ziman, "maximal consensuality"[12]—the democratic rule by jury and consensus. Not everyone is interested, say, in cosmology. If one has a question in cosmology, one accepts as true answers given by men interested in the subject, and their knowledge, in turn, was largely formed by absorbing the knowledge of their similarly minded (interested) ancestors and contemporaries. The ill side effects of specialization that often ensue are well-known[13] and may simply be based on normal adaptation—breathing the same air long enough, one cannot smell it anymore.

The concept of consensuality is, nevertheless, useful, provided we remember cases like that of René Blondlot's fantastic N-rays[14, 15] (accepted by mutual consensus of French authorities) and not forget Francis Bacon's words: "Anticipations [theories] are a ground sufficiently firm for consent; for even if men went mad all after the same fashion, they might agree one with another well enough."[16]

We need not dwell here on this very large topic once named epistemology and now cognitive science; the point is that truth, including scientific truth, is a relative phenomenon. What was true yesterday is false today, true to one, false to others. The validity of a new truth will, therefore, generally depend on how many people are at that moment ready to accept it; and this depends largely on how many are pleased by it—either by the emotional comfort it provides, by its rational elegance, or by its practical utility to society.

When almost all perceive an event in only one particular way, it attains almost absolute certainty. A heated wire emits lights, every one can see it—it is true reality—except in a society of the blind, but then, this society itself is not a true representation of mankind. A true fact of perception is, therefore, related to the established view of man's physiological normalcy.[17] And finally, since realistic concepts can be formed only on the basis of some perceived information, it follows that the veracity of a concept depends on its affinity to truly perceived data. We may, of course, form concepts—like heaven and hell—that are not based on perceived data, but then, their validity can ill be proven and is justly in doubt. In order for a physical fact to be accepted as true, it ought to be perceived as nearly as possible independently of the position where the fact was observed—what is true in New York must be true also in Moscow.[18] True facts of nature ought also be independent of time; the heating effect of fire must have been as true to prehistoric man as it is to us. The assumption is that man's physiology and his perceptual mechanisms did not materially change over time. Therefore, all true facts are reproducible in different places at different times.

REFERENCES

1. Rhinelander, P. H.: Is Man Incomprehensible to Man? San Francisco, W. H. Freeman, 1974.
2. Monod, J.: Chance and Necessity, New York, A. Knopf, 1971, p. 21.
3. Mach, E.: Why has Man two Eyes? in: Popular Scientific Lectures, La Salle Illinois, The Open Court Co., 1943, p. 88.
4. Nisbett, R., Ross, L.: Human Inference. New Jersey, Prentice Hall, 1980.
5. Klein, G. S.: Perception, Motives, and Personality. New York, A. Knopf, 1970.
6. Zuckerman, H.: The Sociology of Nobel Prizes. *Sc. Am.* 217:25, 1967.
7. Buckhout, R.: Eyewitness Testimony. *Sc. Am.* 231 (6):23, 1974.
8. Warren, R. M., Warren, R. P.: Auditory Illusions and Confusions. *Sc. Am.* 223 (6): 30, 1970.
9. Gregory, R. L.: Visual Illusion. *Sc. Am.* 219 (5):66, 1968.
10. Holton, G.: On the Psychology of Scientists, and their Social Concerns, in: The Scientific Imagination, Cambridge Univ., 1978, p. 229.
11. Polanyi, M., Prosch, H.: Meaning. Univ. of Chicago, 1975, p. 182.
12. Ziman, J.: Reliable Knowledge. Cambridge Univ., 1978, p. 6.
13. Ortega y Gasset, J.: Die Barbarei des Spezialistentums, in: Der Aufstand der Massen, Hamburg, Rowohlt, 1956, p. 79. (Also: The Revolt of the Masses, Norton, 1932).
14. Blondlot, R.: "N" Rays, A Collection of Papers Communicated to the Academy of Sciences in Paris. London, Long mans, 1905.
15. Wood, R.: The n-Rays. *Nature* 70:530, 1904.
16. Bacon, F.: Novum Organum, in: Man and the Universe (ed. Commins, S., Linscott, R. N.), New York, Washington Square Press, 1966, p. 85.
17. Ronchi, V.: Optics, the Science of Vision. New York, New York Univ., 1957, p. 9.
18. Franks, F.: Polywater. MIT Press, Cambridge Mass., 1980.

Space

Newer advances in understanding the human body and its various functions—particularly in cognitive neurophysiology and developmental psychology[1-5]—underpin the apparent fact that human cognition is based on neuronal activities. This understanding altered the view, first developed by Immanuel Kant (1724-1804), which saw cognition founded upon certain transcendental concepts beyond human experience.[6] It now appears that the knowledge we are conscious of—as a form of information storage—resides in the cerebral cortex whereto it arrived by means of nerves from special sensory end organs and other parts of the body; other knowledge arrives to subcortical centers and remains largely subconscious whence it may be retrieved, as Freud showed, by an arduous act of search and analysis.[7] This neuronal activity of information retrieval and storage begins in utero before birth, as evidenced, for instance, by the embryo's reactions to sound. The term *knowledge* does not here include rudimentary automatic activities, such as metabolism, which evolved by genetic transmission of chemical compounds.

Aside from data perceived through specific sense organs, the brain is fed proprioceptive information about the position of the body and its extremities. Proprioception is very primitive, remains mostly subconscious, and starts before specialized sense organs attain their proper function. It is prerequisite to normal muscular activity, for in order to activate a muscle, information must be available about its state of contraction or relaxation and the state of contraction of its antagonistic muscle.[8] At the time of birth, man thus already possesses information, first about his own body, concerning positions in space—three-dimensional space.

The concept we form of three-dimensional space, based on perceived sensory data, is present at birth and, yet, is not transcendental; namely, it is not an essential feature that must necessarily be accepted a priori when talking of man and his world. Practically though, no human being has yet been described who lacked—consciously or subconsciously—a concept of three-dimensional space.

"Biological and psychological research combine to confirm the conclusion that, as regards the intuition of space, the nativistic view can all the more be maintained. The chick has scarcely broken from its shell than it is seen to be at home in space and pecking at everything that excites its attention."[9]

Saying that three-dimensional space is common sense and common experience means that the largest possible consensus, a consensus formed by all mankind, perceives it in only one way. This perception of space is augmented after birth by specific information gained through the sense organs—such as the visual perception of nearness and distance, left and right, up and down—information closely tied to that received from the semi-circular canals situated in the inner ear on three different planes, corresponding to the three dimensions. This development of spacial perception after birth had been thoroughly studied by Jean Piaget and his school.[10-13]

Accepting man a priori and recognizing that he arrives in this world with a concept of three-dimensional space, it is yet necessary to describe this space.[14] The prerequisite task of exploring space with the intent of discovering or arranging in it a rational system is based on the need to understand events in it. We need a systematic order amenable to human perception and easy conception that will aid orientation in space prior to taking action in it. In empty space, we know not where we are—nor whether we are coming or going—but with some order, we can find our way and then march on.

Given space and the task of instilling some order in it, we begin with the smallest conceivable building block within space. For a definition to be widely applicable, it must consist of a minimum number of new terms, the aim being to define and explain the maximum number of entities and events by the least number of entities that are beyond definition and comprehension.[15] In addition, a strictly valid definition cannot include the term to be defined or, at least, ought to admit as little of it as possible.

The smallest amount of space is termed *a point*. When we say *a*, we mean *one* and imply that we know that it differs from *two* or any other number. A point is said to have no dimensions—no length, width, or depth—and may thus seem a purely imaginary abstract concept. Inasmuch though as any image—any concept—is based on some perception, the dimensions of a point are related to the size of the space under discussion. In the all the universe, a point may have the dimensions of the sun while within the space of a molecule, a subatomic particle may be seen as a point; every point marked on paper has real three dimensions, albeit very small. James Clerk Maxwell[16] named it "a material particle: A body so small that, for the purposes of our investigation, the distances between its different parts may be neglected." The concept of the point, as the first step on the way toward rationality, stands at the beginning of geometry and other systematic knowledge, and was thus no small feat of the human intellect.

A single point in space does not establish any order, and therefore, we introduce another point. We term the space between the two points *a line*, or unidimensional. The smallest amount of space between two points is given when they are adjacent to one another, and this space we term *a straight line*.

The term *straight* often presumed knowledge of what was crooked. When Euclid[17, 18] conceived of the straight line, he tacitly assumed its existence on a flat plane, but *flat* is a term that may be defined only in relation to a third dimension. Euclid's axiom that the smallest space between any two points was a straight line tacitly presupposed that the position of the points was already fixed on a plane and that the form of the plane was similarly known. These presuppositions (premises) were not written into Euclidean geometry because they were taken for granted as commonsense human experience. It was thus possible in the last century to invent geometries in which the positions of points and the shape of the plane was made to vary—a manifold of n-dimensions and where Euclid's axioms did not all hold; these non-Euclidean geometries were not based on experience and were termed *analytic* as distinct from the *synthetic*.

When the position of two points is given, the line between them consists of an infinite number of points because these are defined as infinitely small. From within the line, no order of magnitude or sequence may be established because no matter what the spatial interval between the delimiting points, the number of points remains infinite whether the line is long or short. Suppose you stand with many other people in a line. All you see is the front or back of one person to your one side and the front or back of a person to your other side. You can form no idea of where the line begins or ends, if at all, or what shape it has; and hence, you can have no idea what position in the line you occupy. If you wish to form an order of magnitude, you may look at yourself and the space you occupy and imagine that ten people to your right ought to equal ten people to your left. But you cannot be certain because all the people to your right may be fat, and all those to the left slim, so that an equal number of each will yet occupy unequal space. Position in line, the shape of a line, and distances within it may only be ascertained when you consider it from outside the line, i.e., from a second dimension.

The elementary branch of mathematics, arithmetic, presupposes the concept of singularity: the one; without me, without man, there is nothing—zero. Zero is assumed to have a fixed position whence the numbers proceed in a given linear direction to the right: 1, 2, 3, and so forth. The position of zero is, however, ambiguous because without *one* there is no entity at all in relation to which a position may be fixed. The definition of zero would have to be expressed as $0 = 1 - 1$ where the negative sign symbolizes the elimination of *one*.

Traditionally, *one* denotes a unit position to the left of zero, presupposing again that zero has a fixed position on a plane from which another position may be established to form a straight line with some direction. Along this unidimensional line on a two-dimensional flat plane, the numbers proceed to the left or right. When zero denotes nothing, negative numbers are meaningless. When zero denotes the starting position, negative numbers denote elimination units, subtraction units in terms of distance; $4 - 2$ means four distance units to the right of zero and two units to the left of zero, which leaves two units to the right.

There is little doubt today that the concept of numbers evolved from real perceived experiences, though Pythagoras and his school were so impressed with the seemingly transcendental power of numbers and their geometrical equivalents that they divined them to form a religion. True believers have existed in every period since.

This leads us to an important concept: the concept of distance. A real point may be sensibly perceived only when it has three real dimensions although these may be chosen as small as the space under study requires. Distance similarly correlates to real perceived information. When we look at two points on paper, the distance between them relates to the space between retinal receptors in the eye that is, in turn, judged by reference to preconceived information about the size of the page or the room. However, one and the same distance in space may occupy different distances on the retinae of different eyes according as their sizes vary. A large eye that may possess per area more numerous retinal elements than a smaller eye, or receives a larger optical image, is able to divide that distance into smaller units (i.e., its visual acuity is better); and it may see distances that are invisible to a smaller eye. A person with one large and one small eye sees a given line longer in one eye than in the other—aniseikonia;[19] his brain must then choose between the two images in order, for instance, to decide how big a step to take for a given distance. One eye is therefore subconsciously chosen as the dominant. At the same time, each eye within its own system can, of course, decide what size is larger than another, but no *absolute* sense of long or short is possible.

There is no distance apart from human perception, and this perception is not an independent entity but exists only in relation to a similar entity, an agreed upon standard. In order to define and determine distance, a frame of reference is prerequisite, and the choice of this frame is completely arbitrary. Traditionally, the frame was chosen from among those systems or objects that appeared the largest and most stable. Each nation had its own distinct system—such as the English yard, the Portuguese covado, or Japanese shaku. To a commission gathered after the French Revolution, the earth seemed large, stable, and convenient enough a frame to which events on it could be referred. The earth's circumference was chosen as constant, and its division into units—the meter—was set by this convention and then established by tradition. There is nothing sacred, transcendental, or universally true about this frame of reference, but a choice had to be made. It was thereafter possible to state unambiguously what was short or long and where on earth was one position compared to another.

The entire science of Euclidean geometry deals with comparisons and congruities: one line is shorter than another, one triangle is incongruent to another, or one volume contains another. The unspelled premise of the science was the definition of distance. This was taken for granted, but may be termed the universal constant of Euclid. It has since become clear that in order to state unambiguously the position of a point or the length of a line, a frame of reference must be given because position and distance are relative terms that exist only in comparison to similar terms. Take, for instance, a sentence from Maxwell: "The position of B relative to A is indicated by the direction and length of the straight line AB drawn from A to B."[20] When he said "position," he already tacitly presupposed a frame of reference and was then able to talk about "direction," "straight," and "length." Properly phrased, the sentence must read, "Within a given and known three-dimensional frame of reference, the position of B," etc. One point in empty space does not constitute a position, and positions and lengths cannot be known in reference to only a single other point.

According to whether we have a frame of reference or not, we can distinguish between position and real position, between a line and a real line, etc. On a real line

ABC, distance AB equals distance BA; the points A and B are equidistant. On a real one-dimensional line, no more than two points can be *mutually* equidistant. The distance BC may equal AB, but all three points—ABC together—are not equidistant because the distance from C to A does not equal its distance to B. When A has a real position, the line has a direction starting with A; and *from this position*, in this direction, the distances are sequential and no two are the same.

Thus based on perceptual reality, we are able to define points and the unidimensional distance between them. In order to widen our concept of space, we now introduce a third point not in line with the other two. The space between these three points is termed a *plane*, or two-dimensional. For any chosen minimum unit of unidimensional (linear) distance, the smallest two-dimensional space is an equiangular (equilateral) triangle. From among any three points, one distance, say

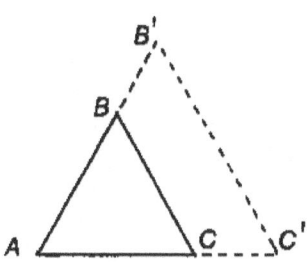

AB, may be set as standard; compared to this frame of reference, AC or B'C' are longer. But in a system consisting of only three points not in line, the position of any one of them cannot always be unambiguously defined because at a minimum, the three form an equilateral triangle in which no preferred distance is discernible to which other distances may be referred. In order to decide with certainty what is longer or shorter, we need a fixed standard for comparison, and three points between themselves do not furnish such a standard because they may each be equally distant from one another.

When position B (figure) is changed to B' while C is equidistantly changed to C', it is impossible to state whether A, B, or C changed their position. On a plane, no more than three points can be mutually equidistant. Points A, B, and C may be equidistant and also points A, B', and C, but not points A, B, C, and B' together because AB is not equal to AB'.

We then add a fourth point not on a plane described by the other three. The space between the four points is termed a *solid*, or three-dimensional, and for any minimum unit of distance, the smallest three-dimensional space is a tetrahedron of equiangular triangles. This tetrahedron does not provide a frame of reference for its constituent parts, i.e., any of the four points may occupy equidistant positions relative to the others without means of ascertaining from which one of them is the position to be judged. In three-dimensional space, no more than four points can be mutually equidistant. When a fifth point exists, its distance to one of the other four must differ from all other distances. One may, in this manner, choose a preferred position from whence all others may be compared. Once a preferred position is chosen, one may state in reference to it what is long and what is short. In this three-dimensional system, no two real points may occupy the same position.

The four points, defining three lines, constitute a system of coordinates named after its inventor René Descartes who, for the first time, systematized the need for some frame of reference. The space described by this system is termed Euclidean space, or real space. There seems to be no magnitude, physical or other, without a frame of reference; in itself, the symbol D for length, V for velocity, or T for temperature has no real meaning. The assumption that a frame of reference was an artificial product of pure human imagination and reasoning, without factual basis, allowed some metaphysicians (mathematicians) great liberties, and when these were then turned around and applied to the real world of facts, they often led to inconsistencies. Concerning real space, Henri Poincaré said, "The language of three dimensions seems the best suited to the description of our world, even though that description may be made, in case of necessity, in another idiom."[21]

The other idiom Poincaré referred to was the non-Euclidean geometries, first invented by Carl Friedrich Gauss (1777-1855) who pointed out that Descartes's frame of reference need not be the only one. These geometries were further developed by N. I. Lobachevsky (hyperbolic geometry), G. F. B. Riemann (elliptic geometry), E. Beltrami, J. Bolyai, and others. They sprang originally from Euclid's parallel axiom and demonstrated the logical coherence of systems other than Euclid's three-dimensional one. The intellectual capacity to conceive these spaces did not deny the existence of the real three dimensions, but rather seemed to add to it. Over the centuries, man has held many conceptual systems, completely valid and logically consistent within themselves; whereby, their real veracity and utility depended on their affinity to perceptual data. The term that distinguished Euclidean space from all others was *distance*.

In geometry, and mathematics in general, it is admissible to premise one dimension or coordinate as an a priori constant. There is no question as to the reality of the coordinate—it is presupposed as given without requiring proof of its physical and perceptual existence. Instead of Euclid's three dimensions, some constructed a two-dimensional system, for instance, such as would be formed by bending a sheet of paper unto some well-defined form (a bent plane must of course be in three dimensions). On such pseudospherical surfaces, various axioms (such as parallel lines) were proven false, or different, which previously appeared immutable according to Euclid. The new space—like the inner surface of a sphere—was unlimited (had no beginning and no end) but finite (of certain area). With the admission of certain propositions and axioms, the Euclidean and non-Euclidean geometries validly (logically) existed side by side.

When the non-Euclidean geometries became better understood and more widely disseminated toward the end of the last century, it caused noticeable disturbance among natural philosophers: if it was possible to construct a geometry where Euclid's axioms did not all hold, then these axioms could not be transcendental—taken a priori as true. And since understanding of physical reality was based on geometrical and similar axioms, and on the logical constructions that deductively followed from them, their mutability cast doubt on the validity of the established conception of the physical world. And yet, to quote Helmholtz,

Land surveying as well as Architecture, Mechanical Engineering as well as Mathematical Physics, are all constantly computing the most varied spatial relationships according to Euclidean geometrical laws; they expect the success of their experiments and constructions to follow these computations, and there is yet no known case in which they were disappointed, provided they computed with correct and sufficient data.[22]

Elements of logic teach us that *logically*, something may be absolutely *valid* but *in reality* quite *untrue*; there is a fundamental difference between logical validity and real truth. The difference manifests itself in the relationship between conceptual ideas (hypotheses) and perceptual (experimental) data or between theory and practice. Unless there is a flaw in the logical construction—in the computer—a theory is always valid. Its real truth may only be tested by checking first its premises (input) and then its conclusions (output).

Various logical systems, such as Euclidean and non-Euclidean geometries, may validly coexist simultaneously side by side.[23] In the physical realm, such a state of affairs is unacceptable. The premises and presuppositions of one system, say chemistry or physics, must coincide with those of all others, such as physiology or biology. No two systems that differ in their presuppositions and their conclusions will exist simultaneously for very long. A certain grand order in nature consistent in itself is always tacitly assumed, and it is the business of science to discover it. Understanding nature means feeling in unison with the real external world—it is then less mysterious and threatening, more friendly and predictable.

The theoretical investigation of the mathematical possibilities above referred to [non-Euclidean geometries], had, primarily, nothing to do with the question whether things really exist which correspond to these possibilities; and we must not hold mathematicians responsible for the popular absurdities which their investigations have given rise to. The space of sight and touch is three-dimensional; that no one ever yet doubted.[24]

In the real physical world, the premises, the coordinates, and the universal constancy of any entity cannot be a priori assumed without experimental—perceptual—proof; and this proof, for its part, is continually changing by ever-increasing number of empirical data acquired by evermore sophisticated tools. A physical system founded on old coordinates or universal constants will not remain true when new data contradict them. In contrast to any number of logical (mathematical) systems that are each consistent, the one physical system—our entire concept of the real world—must be altered when inconsistency arises since we cannot admit of the existence of two different worlds for the one man.[25] When the shortest physical distance between two points is x meters for Tom, it must also be so for Dick. If Harry says that according to his system, the shortest distance is y meters, then Harry is not a human being equivalent to Tom and Dick—although he may certainly, and without any inconsistency at all, be a pseudospherical non-Euclidean entity who grows shorter and younger as he swiftly moves along and vanishes into his finite but unlimited horizon.[26, 27]

REFERENCES

1. Chusid, J. G.: Neuroanatomy and Functional Neurology, 14th ed., Los Altos, Lange Medical, 1970, p. 200.
2. Fields, W. S., Abbott, W. (ed): Information Storage and Neurological Control, Springfield, C. C. Thomas, 1963.
3. Carterette, E. C., Friedman, M. P. (ed.): Handbook of Perception, New York, Academic, 1973.
4. Wooldridge, D. E.: Sensory Processing in the Brain, New York, Wiley, 1979.
5. Gazzaniga M. S. (ed.) The Cognitive Neurosciences III. 2005; Cambridge, MIT Press
6. Kant, I.: Kritik der praktischen Vernunft, Riga, 1788.
7. Freud, S.: Collected Papers, London, Hogarth, 1950.
8. Harris, D. A., Henneman, E.: Feedback Signals from Muscle and their Efferent Control, in: Medical Physiology, 4th ed., (V. B. Mountcastle, ed.), St. Louis, Mosby, 1980, p. 702.
9. Mach, E.: The Analysis of Sensations, New York, Dover, 1959, p. 126.
10. Piaget, J.: The Child's Conception of Geometry, New York, Basic Books, 1960.
11. Piaget, J.: The Mechanisms of Perception, New York, Basic Books, 1969.
12. Piaget, J.: Biology and Knowledge; an essay on the relation between organic regulations and cognitive processes. Univ. of Chicago, 1971.
13. Anderson, J. R.: Cognitive Psychology and Its Implications. San Francisco, W. H. Freeman, 1980.
14. Jammer, M.: Concepts of Space. Cambridge, Harvard Univ., 1954.
15. Hibben, J. G.: Logic. New York, Scribner, 1923, p. 44.
16. Maxwell, J. C.: Matter and Motion. New York, Dover, 1953, p. 3.
17. Euclid: The Elements. London, J. Daye, 1570.
18. Forder, H. G.: The Foundations of Euclidean Geometry. New York, Dover, 1958.
19. Burian, H. M., Noorden, G. K.: Binocular Vision and Ocular Motility. St. Louis, C. V. Mosby, 1974, p. 136.
20. Maxwell, J. C.: op. cit. p. 4.
21. Poincare, H.: Science and Method. New York, Dover, 1952, p. 113.
22. Helmholtz, H. F.: Vortraege und Reden, 4th ed, Braunschweig, Vieweg, 1896, vol. 2, p. 1. (also: Popular Lectures on Scientific Subjects, New York, Appleton, 1873).

23. The parallel Postulate. The Dividing Line Between Euklidean and Non-Euklidean Geometry, and what the latter means. *Sci. Am.* 1920; 123: 565.

24. Mach, E.: Science of Mechanics, Chicago, The Open Court, 1942, p. 591.

25. Tisza L.: Can we learn from history? In: revisiting the foundations of relativistic physics. Dordrecht, Kluwer Academic Publ. 2003: 555-568.

26. Einstein A. Geometry and Experience. In; Sidelights on Relativity. London; Methuen & Co; 1922: 27-56.

27. Mach E. The Science of Mechanics. Op cit pp: 271-305.

Time

The definition of time has always posed a great many difficulties. Time, according to Aristotle, was infinite and, therefore, could not be defined. "What, then, is time?" inquired St. Augustine.[1] "If no one asks me, I know what it is. If I wish to explain it to him who asks me, I do not know." What is time? In order to define anything, it must relate to something else or the term must be described in other terms, but time seems an entity standing alone and of its own. Kant, for example, believed that time was a purely subjective condition of our intuition—it was axiomatic, transcendental. Intuition means subconscious knowledge, and since all knowledge—all information and cognition—must originate somewhere, intuitive subliminal information must also reside concretely in some form, perhaps in the guanine or the adenine molecule of DNA. That our knowledge of time is derived from some form or shape of molecules has not yet been established.

In the absence of guidelines from this direction, we must seek a meaning for time elsewhere, for if we are to operate with the term *time*, we need a workable definition, or to cite Albert Einstein, "Physicists have been obliged by the facts to bring down from the Olympus of the a priori our concepts of time and space in order to adjust them and put them in a serviceable condition."[2] If anything—a concrete object or an abstract term—is to be used rationally, it must be known and defined. The definition may change as the human intellect evolves, but a provisional definition is useful and, therefore, preferable to none.

The predicament that attended a definition is perhaps best illustrated by standard dictionaries. For instance, Webster defined time as "the measured or measurable period during which an action, process, or condition exists or continues." This definition, as many similar ones, breaks a cardinal rule of logic, which states that a definition must not contain a term equivalent to the term to be defined. What is a period? Webster said, "A portion of time," "an interval of time." Substituting this sentence into his original definition of time, we obtain the following: "Time is a portion of Time," which makes little sense. Another definition in Webster states, "The point or period when something occurs." The definition of *when* is given as, "At what time" or "At or during which time." Hence, by substituting, "Time is a point or period at what time, or during which time, something occurs," which is, again, obscure. The error illustrated

170

by the dictionary is termed a circle in the definition (*circulus in definiendo*) and leads, as all circles do, back to the starting point with no advancement of knowledge.[3]

Isaac Newton was hesitant to furnish definitions: "I do not define time, space, place, and motion, as being well known to all."[4] But he, nevertheless, proceeded in order to distinguish between definitions meant for philosophers and those for common folk. Newton's true definition of time was "Absolute, true, and mathematical time, of itself, and from its own nature, flows equably without relation to anything external, and by another name is called duration." The question is, what did he mean by the term *flow* and the term *external* in relation to time? A river flows when its waters change position; what does it mean when time flows? A tree is external to a house; what is external to time? The ambiguity of Newton's definitions did not compromise the further development of his thesis, for the state of knowledge did not then require a more accurate definition. This need became acute only after Henri Becquerel, one morning in Paris in 1896, discovered that the photographic plates in his drawer bleached[5]—a discovery that eventually led to Hiroshima.

Over a century ago, it was said,

> It might appear possible to overcome all the difficulties attending the definition of "time" by substituting "the position of the small hand of my watch" for "time." And in fact such a definition is satisfactory when we are concerned with defining time exclusively for the place where the watch is located.[6]

This definition, regrettably, also seems unsatisfactory, even for the place in which the watch was located because we have no definition of a *watch*. A definition cannot be in more obscure language than the term to be defined. If time is the position of the hand of the watch, we must know what determines this position. Looking at the position of mercury in a thermometer provides as little enlightenment concerning thermodynamics and the nature of heat, at the place where the thermometer is located, as looking at a yardstick would provide on the nature of space. Measuring devices presuppose some idea of what is to be measured, or as Voltaire once said, "*l'horloge implique l'horloger*." No matter how long and arduously we investigate a watch, all we shall find is screws, springs, and spinning wheels but are not likely to arrive at a definition or understanding of the entity *time*.

The apparently insurmountable obstacle on the way to forming a definition of time may have been the absence of a clear frame of reference: clear terms with which to form the definition. What terms may we use to form a valid definition of time? We can only use what we have, and what we have so far is man in his three-dimensional Euclidean space. We are, therefore, constrained to define time with these facts that means, at present, time is not an independent fact. The first fact—man—is axiomatic and the second—space—is a fact of perception. Thus far, no independent perceptual mechanisms meant to sense time are known.

Given at least one-dimensional space, I offer here my definition:

Time is the only entity present with one point in different positions.

A point can have no more than one and the same position at the same time. Whenever in space a point occupies two or more positions, time exists; when nothing in space changes position, there is no time, or in modern words, time is frozen.

The definition appears to me valid on the following grounds:

1. It is not circular; it does not contain terms equivalent to the term to be defined. When we dealt with position in space, we noted that any position, and change in position, exists only within a given frame of reference; there is no absolute position. Consequently, time too is always referred. The meaning of this relativity is clarified when we retrace our steps to the concept of man and his three-dimensional space.
2. Our definition is not too wide or too narrow; it deals with a fundamental concept, and its width is proportional to the task.
3. The definition is *per genus* (a point) *et differentiam* (in different positions).
4. It is precise and clear; the only entity. The universe contains, of course, innumerable other entities, but in an attempt to understand any of them, we may begin with the smallest common denominators, that is, points in space.
5. It is not formed by negative attributes.

But even if we are perhaps satisfied that this definition of time meets with the criteria of logic, we must not rest here; for time is also a physical entity and, as such, must conform to the real physical world. What is the reality of time?

We know that position—and hence also a change in position—is always related to an arbitrary frame of reference; the larger and more stable the latter, the better. Since the dawn of civilization in Babylon, the arbitrary frame for time has been the sun—or another large celestial body—to which a change in the earth's position was referred (or vice versa when the Ptolemaic system prevailed). If one day the earth should stop changing its position relative to the sun, there will be no solar time; all solar clocks (dials) will have stopped, and the solar day will last forever. If the circumference of the earth's orbit should double, the length of the solar year will double. Nowadays, time is occasionally reckoned by atomic clocks—the frame of reference being the rate of radioactive decay, which means the change in the position of atomic and nuclear particles. Should the number of positions and their distances suddenly decline to half their original value, the duration of the atomic hour will change commensurately.

While Babylonians reckoned their time relative to the sun, the Israelites set their time relative to the position of the moon. Babylonian and Israeli times—the duration of the month—did not coincide because the positions were incongruous, and any specific position within the calendar, any date, did not occur at the same time; or in Greek, they were not synchronous, or in Latin, not simultaneous. Jewish holidays continually fall on different dates according to any other nonmoon calendar.

On earth itself, the position of longitude has been set, arbitrarily and by convention, to be that meridian that passes through Greenwich, England; and clocks were synchronized according to Greenwich Mean Time. There is nothing sacred or innately true about this measure of time, but it had to be done if some order was desired, similar to the necessity of choosing an arbitrary unit of length.

A zero time interval on solar clocks exists when the earth ceases its spin and orbital rotation around the sun. Zero time on atomic clocks exists when radioactivity stops; total zero time exists when absolutely nothing changes its position. Should all possible positions in the entire universe remain stable, no time whatsoever would elapse; when all electrons cease spinning and all celestial bodies stand still, time will not be there—the temporal interval will be zero. A man will not grow older in whose body has stopped all atomic and molecular changes that determine metabolic and neuronal activity as long as this state of immobility remains. Life under these circumstances will be frozen.

Time positively exists when any position is changed—it is zero when no position whatsoever changes. Nothing is more immobile than stationary, more fixed than at rest. It follows that negative time is impossible; time flows in only one temporal *direction*.[7,][8] As soon as something moves, time elapses; if it is stationary, time is zero. One may theoretically remain frozen at any given age, but sadly, one can never grow younger.

The unidirectional flow of time sets a limit to determinism and predictability. Since time is not cyclical, no event is ever the same, just as no human being is ever the same as any other. The predictive value of scientific laws on which Laplace placed such great confidence is somewhat limited on this account. One occasionally reads that in a Newtonian system, time was perfectly reversible or that astronomical time was reversible.[9] This depends on the frame of reference. For instance, in the long run, the sun loses its mass by constant radiation; and the orbits of the planets and earth cannot, therefore, remain unaltered. Thus solar time must change—the days will be longer but not run backward.

The question sometimes arises as to the meaning of the term *the present*. The present moment was said to be between the eternity of the past and the eternity of the future. The next question that must then follow is, what is eternity? Eternity, according to our definition, is that state in which time does not exist. Since time exists only with change in position, eternity will begin tomorrow if tomorrow all positions in the universe will be stationary. Opponents of Darwin's evolutionary theory argued that it was possible that the world was created yesterday; it was created with everything already in it, including our archives and memories of what we believed has happened last year. There is no way of contradicting them—their logic was valid, but of dubious truth and utility.

Eternity may indeed exist at any moment past or future, unless we assume that time always *flows equably*, which is to say the *sum total* of all changes in positions in the universe is always constant. As expressed by Newton, the assumption that "absolute, true, and mathematical time, of itself, and from its own nature flows equably" implies a general principle of motion; namely, the sum total of all motions is forever constant. Were it not, time would not flow equably, at a uniform rate.

Suppose—for the purpose of illustration—that in the entire universe, only six points exist, six atoms, or six stars. Four of them, not on one plane, are stationary relative to one another and form the frame of reference. A fifth point A changes position a distance x referred to the frame. Time is determined by this point A and is supposed to flow equably, i.e., the rate of change in the positions of A is continuous and uniform. Now take a sixth point B, which covers distance y. Since we suppose that A moves uniformly, B's rate cannot change because if it increased, for instance, the rate of motion of A referred to B necessarily decreases, which cannot be when we assume that the motion of A is uniformly constant and determines the time. The time of B would then have to be decreased, i.e., it dilates, which is incompatible with the assumption that it flows equably. It is, of course, then possible to change the frame of reference and refer time to point B, but then A cannot change its rate if B is assumed to flow uniformly. If there is only one time, and this time flows uniformly, the sum of position changes of A and B must remain constant.

It is not our aim here to go into dynamics, but when we define energy as the cause and effect of change in position, it becomes immediately evident that the concept of time flowing equably (the sum total of all motions being constant) leads to the laws of thermodynamics dealing with the conservation of energy, or the laws of thermodynamics presuppose uniform flow of time.

One last concept in connection with time that occasionally leads to some ambiguities is termed *instant* or *instantaneity*. The talk is often about "instant event" or "instant action." When the implied definition of instant is "a very short interval of time," then the expressions make sense; when the definition is "at some point in time," it is false. No matter how swift the event, how fast the motion, it cannot exist without the lapse of time. At a point in time, there are no events and no motions—only stationary positions. When a mathematician said "We may subject the axes x, y, z, at $t = 0$ to any rotation we choose," he may wish to explain how the rotation ensued without the passage of time ($t = 0$), otherwise his mathematical model of reality will not truly apply.[10]

"The speed of the train at 12 o'clock" is meaningless because at any given point in time, the train may only have a position; "it was in Greenwich." The proper statement would be "the speed of the train between 11 and 12 o'clock."

To cite another example from Maxwell, "Thus when we say that at a given instant, say one second after a body has begun to fall, its velocity is 980 cm/sec, we mean that if the velocity of a particle were constant and equal to that of the falling body at the given instant, it would describe 980 centimeters in a second."[11] In fact, however, one second after a body has begun to fall, it had no velocity at all, only a position. As the distance and time period of measurement is reduced, the motion approaches a stationary position and becomes more and more uncertain and indeterminate. Maxwell added, "The ideas which are suggested to our minds by considering the motions of a particle are those which Newton made use of in his method of Fluxions."

Operations of infinitesimal calculus aim at ascertaining an expression for velocity, for instance, when the distance covered approaches zero (dx - 0). When two positions approach unity, the time needed to go from one to another approaches zero, and commensurately, our knowledge of the velocity tends to zero, or our ignorance tends to infinity. The result may be termed *ignorance principle* or, in deference to common reluctance to admit ignorance, *uncertainty principle*. Newton originally invented calculus to help describe planetary kinetics in the large solar system. Applying his method to swift motions of electrons in a molecule may be as difficult, in spatial terms, as locating the position of a room on a map of the solar system.[13]

REFERENCES

1. Park, D.: Roots of Time in the Physical World. Amherst, Univ. of Mass., 1980.
2. Einstein, A.: The Meaning of Relativity, 5th ed. Princeton Univ., 1956, p. 2.
3. Luce, A. A.: Logic. London, The English Univ., 1958, p. 27.
4. Newton, I.: Mathematical Principles of Natural Philosophy. Berkley, Univ. of California, 1966, p. 6.
5. Becquerel, H.: Sur les radiations emises par phosphorescence. *Compt. Rend.* 122:420, 1896. and 122:501, 1896.
6. Einstein, A.: Zur Elektrodynamik bewegter Koerper. *Ann. d. Phys.* 17:893, 1905.
7. Sachs, R. G.: Can the Direction of Flow of Time be Deter mined. *Science* 140:1284, 1963.
8. Overseth, O. E.: Experiments in Time Reversal. *Sc. Am.* 221:88, Oct. 1969.
9. Wiener, N.: Cybernetics, 2nd ed., Cambridge, Mass, MIT Press, 1961, p. 37.
10. Minkowski, H.: Space and Time, in: Lorenz, Einstein et al.: The Principle of Relativity, New York, Dover, 1952, p. 77.
11. Maxwell, J. C.: Matter and Motion. New York, Dover, 1952, p. 20.
12. Sutton, O. G.: Mathematics in Action. London, G. Bell & Sons, 1958, p. 26.
13. Ridderbos K (ed.). Time. Darwin College Lecture Series (4) 2002. Cambridge Univ.

Motion

It is a property of rest, that bodies really at rest do rest in respect to one another. And therefore as it is possible, that in the remote regions of the fixed stars, or perhaps far beyond them, there may be some body absolutely at rest; but impossible to know, from the position of bodies to one another in our regions, whether any of these do keep the same position to that remote body, it follows that absolute rest cannot be determined from the position of bodies in our regions.

This is from Newton's *Principia*. Position means a spatial relation to another position. A single point in an empty space has no position, neither do have two points. A position is determinable only after four points forming a solid are given. In reference to these points, a fifth point may have only one position. When one point has two positions, the event is termed motion, and it entails a quantity of time. The motion of one point in real space is determinable given a frame of reference for time. The motion of a second point may then be referred to the first in terms of time; it may be faster or slower than the motion of the first point.

The positions of all four points that form an equiangular tetrahedron may change equally—they may all approach or recede from one another at equal rates—without knowing which point moved; only when a fifth point is in motion can its rate of motion be determined with certainty, determined in terms of position and rate in reference to the other moving or stationary four. When we see four or more stars in the sky, we and they cannot all change positions equally—recede or approach at equal rates or equally accelerated rates—without means of choosing first a preferred position to detect these motions.

For practical reasons, the frame of reference is chosen according to apparent size and immobility; an object that appears very much larger and stabile than the object under study forms the frame to which the study refers. An entomologist on a ship who studies the motions of an ant crawling on a table will choose the table as the frame of reference. His science will be little advanced by choosing the ship's port of departure as reference; for whether the ant moves from New York to London or elsewhere is inconsequential to the ant and, therefore, to the study. These points on earth are, of course, the frame of reference for the captain navigating the ship who, for his part, cares little about the motion of the earth in reference to, say, the Andromeda Galaxy.

Given a Cartesian frame, and means for determining time, the motion of point A referred to point B equals the motion of point B referred to point A. The choice of naming the motion is therefore arbitrary, but a choice must be made, and once made, must not change in midcourse if the aim is rational systematic order. When we speak here of motion, we mean linear motion—in a straight line—whereby a straight line was defined as two adjacent points. When a third point is adjacent to one of the two, it may be in any direction, up or down, left or right. In order for three or more points to be in line, the plane must be flat; and this flatness is determined in reference to a solid, i.e., a Cartesian frame. By any scale, the distance between two real points is zero when they touch one another. When one moves ever so little away, the minimum distance between the points will be given by the least number of the smallest units of distance, which will thus form a straight line. A segment of any line of whatever form, when sufficiently reduced, yields a straight line. Before a point can move any other way, the very first and infinitesimally small move will be straight. For such a move, time almost ceases to exist and the term *motion* loses its meaning.

The minimum distance between two points is given when they are on a real straight line. Time is minimum when, for a given distance on the line, the rate of change in position—the velocity—is maximum. Therefore, the higher the velocity, the nearer to a straight line the form of the motion, or anything moving fast will tend to do it in a straight line, thus minimizing the elapsed time.

This conclusion seems fair enough as a guiding metaphysical principle and relates to ideas named the principle of least distance, the principle of least time, and Newton's principle of inertia. Fallacies ensued only when the principle was applied to events where the path of motion was not straight, as Hero of Alexandria did in connection with reflection, or when the rate of motion was not uniform, as Fermat did in connection with refraction, which cases logicians designate as *Secundum quid*—"one of the subtlest and commonest sources of error,"[2,3] And we shall return to the problem in the optokinetic part.

Dealing with motion, we touch upon an idea concerning which Eddington said, "Probably if we could understand it we should not think it so fundamental."[4] No one wins a prize who propounds an idea so lucidly as to render it self-evident. The idea was thought fundamental and, based on it some philosophers, pronounced the death of determinism and causality.[5,6] To grasp the idea in historical context, we need know about momentum. Momentum—or inertia, according to Galileo and Newton—is a feature of a body *in motion*; it tends to continue moving (in a straight line). When hit by a ball, one knows it had momentum. In 1927, Werner Heisenberg seemed to have proven that one cannot accurately determine the position and momentum of an electron at the same time.[7]

Sir William Dampier explained the consequences: "We cannot define to an electron the position in space at a given time, or follow it in its orbit, and consequently we have no right to assume that Bohr's planetary orbits exist."[8] To change position between two points that are so near as to approach one position requires so little time as none. Velocity under these circumstances is indeterminable; the terms *position* (stationary) and *velocity* (a change in position) are mutually exclusive. One can lie in bed and be walking at the same time, but only in a dream.

Light has momentum, i.e., when it hits an electron, the electron's path changes.[11] To know anything visually entails the interaction of light with the electrons of matter, both when the perception is initiated—as when light hits this page and at the receiving end—when this light then acts on the eye or camera. But since light changes the electrons of the object seen, we can never be sure; so said Heisenberg that what we visualize as reality is really true—the act of perception changed the reality.

This said, despair was probably premature unless we insist on carrying metaphysical arguments to absurdity. When we wish to know a pie, its taste, we must eat it; and then there is no more pie to know. The act of perception—in this case, the perception of taste—destroyed the reality. We can never be absolutely certain that the next bite will be the same. But do we need this degree of certainty? The fact that time flows equably forward already guarantees without any doubt that no two events will ever be the same. It seems all a matter of statistical probabilities where limits are set by the required degree of certainty or security, which in turn is set by our state of uncertainty or insecurity.

What were the limits of Heisenberg's certainty? According to undulatory concepts, the visual determination of position and distance—and hence, the motion's path—was limited by the length of a lightwave: "Die hoechste erreichbare Genauigkeit der Ortsbestimmung ist hier im wesentlichen durch die Wellenlaenge des benutzten Lichtes gegeben." When light was believed to consist of waves with a length of about 500 nanometers, it was impossible to make visible anything smaller. In addition, when it was assumed that the velocity of light, or anything else, had an upper limit of 300,000 km/sec, no event could be known that occurred more often than 300,000 km/500 nm times in a second (six hundred million millions). Moreover, in this domain, calculus and differential equations cannot apply; and one must resort to statistical methods of probability, i.e., thousands of observations in Wilson's bubble chamber.

Employing statistics does not necessarily mean playing dice. Medicine in recent times has not fared badly with this method; when penicillin works truly, only ninety-five times out of one hundred the conclusion is that it truly works. Few doubt that the cause of cure of the ninety-five patients was determined by penicillin and who hence refuse it when indicated. Those who are unhappy, unless 100 percent results are guaranteed, have a problem—real as well as metaphysical.

Heisenberg fairly assumed that the known facts of atomic physics supported the undulatory hypothesis of quantum mechanics, and in turn, therefore, reality—our entire concept of the real world—must accommodate to this hypothesis: "The necessity to revise the terms of Kinematics and Mechanics arises immediately from the basic equations of quantum-mechanics." As we shall see later, such Procrustean accommodation of physical reality to mental mathematical precepts led more than once to fabulously odd results.

Where the high magnitude of velocity renders its measurement difficult, two options aside from statistical probabilities are available: either slow down the time—reduce the rate of motion—or lengthen the distance, or both.

1. As in a cinematic film in slow motion, the positions may be more easily detected when the object under observation moves slowly. Once the positions are known, we can reinstitute the original time by multiplying with the reduction factor. In optics, this method applies to light traveling in transparent media.

2. No matter how great the velocity, it may be ascertained when the distance is sufficiently long; when the change in position requires too little time to be accurately measured, we may separate these positions further. Concerning light, this method has the advantage over the first one in its independence upon the possibly unknown properties of the material media. It is used by conducting the measurement over known astronomical distances. The options of advancing beyond uncertainty along these two avenues give optics an advantage over atomic physics, which appears greatly handicapped by the short distances and great velocities, and the limited control over both.

Traditionally, motion and velocity are designated by the letter V or v, distance by the letter D or d, and time by the letter T or t, whereby velocity is $V = D/T$. When inside a car, a ball (A) moves with velocity (v), a distance (d) referred to the car of length (d), and the car itself moves with velocity (v) referred to the road, then the total distance referred to the road, which

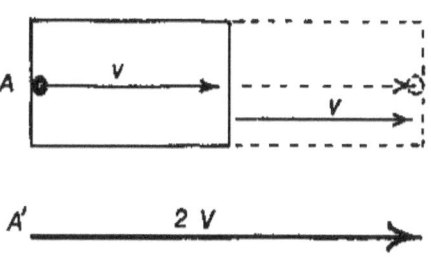

the ball covers in unit time, is 2d, and its total velocity referred to the road is hence 2v. To an observer A' in a passing car moving in the same direction with velocity 2v on the road, the ball will appear stationary. This is perceived reality. We began elucidating it by noting our perception of three-dimensional space, noting the entity of time referred to motion in this space, and we thus arrived at some general understanding of uniform linear motion.

Let us now see what happens when we proceed in reverse as was done by FitzGerald and Lorentz; that is, we presuppose that some uniform motion, say that of the earth or that of light, is the starting point. All we know is this motion that is presupposed to be a universal constant no matter what the motion of the frame of reference or of the observer. We now attempt to form an idea about space, time, and other motions in reference to this motion.[9, 10] Logicians would naturally have a problem with such a system because it breaks one of their cardinal rules of logic that states that a whole class—in this case the class of all motions—cannot be defined by the characteristics of only one member. Metaphysicians may have difficulty with the system's presupposition because at the beginning the universally constant motion that forms the frame of reference must have been measured, at least once, in reference to some fixed yardsticks—be they kilometers or miles—and its duration timed on some clocks, all of which must be supposed to be universally at rest or of known invariable size and rate in some frame of reference that is assumed to be at rest.

Consider the circumstances first according to normal standards (which begin with space and time) and (1) suppose entity A'B' (lower bar) two meters long and stationary while AB (upper bar) also two meters long and moving to the right at, say, 2 m/sec. In one second, A will be above B'; the positions are congruent relative to the abscissa of the frame. (2) When A'B' also moves to the right, but at velocity, say, of 1 m/sec while AB moved at 2 m/sec, then after one second, the position of A will coincide with the midpoint C' on A'B'.

Suppose now (3) that the motion of AB referred to A'B' is constant, whether A'B' is in motion or at rest [(1) = (2)]. AB's velocity is 2 m/sec and A'B' is 1 m/sec in the same direction. Since the velocity of A'B' referred to AB is the same whether A'B' moves (2) or is at rest (1), the position A must be congruent to B after only one second (1). The moving observer on A'B', reasoning in traditional terms, believes that in one second, the position is congruent to C' (2); he must therefore conclude

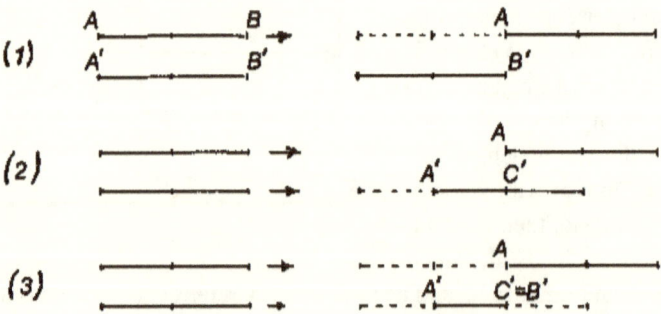

that the distance A'B' contracted to A'C' (3), $\overline{A'C'} = \overline{A'B'}$, two meters became one. In addition, the congruence A over C' occur in one second (3) whereas the observer at (1) measured it as 1/2 seconds, i.e., time dilated from half to one. The opposite occurs when the motion of A'B' is in the opposite direction to that of AB, i.e., distances dilate and time contracts.

What this system required of us, again, was an accommodation of our real perceptual data to the conceptual idea. The accommodation was actually painless when the velocity that formed the frame was in the order of 1,000,000,000 km/hr because our usual motions compared to this speed are infinitesimal and, correspondingly, the deduced changes in our body, or in the entire world, would be fortunately inconsequential. A logical concept that presupposes a form of motion independent of the motion of the frame of reference may be perfectly coherent and consistent within itself. We do not opt for it in this book for the same reason we overlook the Ptolemaic solar system—it is too complicated, i.e., its conclusions are hard to perceive. Other reasons concerning the system's presuppositions are exposed in the optokinematic chapter.

So far, we have treated uniform linear motion. We noted that the least quantity of motion is given when the change in position is smallest, and we defined energy as the cause and effect of motion. Before a point may move anywhere else, it moves to the adjacent position, and before it moves at all, it must be changed from a position of rest. Any change from a state of rest to a state of motion is termed *acceleration* and requires the presence of energy as a cause. When a state of motion returns to a state of rest—an event termed *deceleration*—energy exists as the effect.

When energy as cause stops after a point reached a certain value of motion—a velocity of some magnitude—the point will continue, as Galileo and Newton had shown, with the same rate of linear motion, and this is termed *inertia*. With no more energy, the motion remains uniform and linear. When energy is added, it may vary the motion in two ways: (1) Change its rate—either increase or decrease it in the same direction—or (2) change its direction. When the direction of a linear motion changes, it is either caused by the addition of energy or effects loss of energy. In the absence of a source of energy, any change in direction entails reduction in the value of the motion (lower velocity), a fact that plays a cardinal role in optokinetics.

REFERENCES

1. Copernicus,.: On the Revolutions of the Celestial Spheres. In: The Philosophers of Science (Commins, S., Linscott, R. N., ed.), New York, Washington Square Press, 1966, p. 61.

2. Luce, A. A.: Logic. London, The English Univ. Press, 1958, p. 165.

3. Hibben, J. G.: Logic. New York, Scribner's, 1923, p. 161.

4. Eddington, A.: The Nature of the Physical World. Ann Arbor, Univ. of Michigan, 1958, p. 207, 220.

5. Bergman, H.: Der Kampf umdas Kausalgesetz in der juengsten Physik. Braunschweig, 1929, p. 39.

6. Cassier, E.: Determinism and Indeterminism in Modern Physics; Historical and Systematic Studies of the Problem of Causality. New Haven, Yale Univ., 1956.

7. Heisenberg, W.: Ueber den anschaulichen Inhalt der quanten-theoretischen Kinematik und Mechanik. *Zeitschrft. f. Phys*. 43:172-198, 1927.
8. Dampier, Wm. C: A History of Science. Cambridge Univ., 1961, p. 396.
9. Einstein, A.: Ueber einen die Erzeugung und Verwandlung des Lichtes betreffenden heuristischen Gesichtspunkt. *Ann. d. Phys*. 17:132, 1905.
10. Lorentz HA. In Einstein's: Principle of Relativity; pp 3-34. New York, Dover, 1923.
11. Minogin VG, Letokhov VS. Laser light pressure on atoms. 1987, New York, Gordon and Breach Science Publishers.

Light

The term *light* ordinarily denotes a fact of perception sensed specifically by the eyes of all normal mankind. Historically, in forming a mental concept of light—in trying to define it in other terms—it was natural to do so in terms of concrete objects amenable to perception by means other than light, such as touch; light was regarded a material body, something of substance that impressed on the mind a familiar experience. The main problem was to explain (conceptualize) how the seen objects acted on the eye, an early encounter with the problem of action at a distance, which resurfaced much later in connection with gravity.

Early concepts about the nature of light were intimately and naturally tied with interest in vision and blindness. Optics is Greek for the science of vision, whereas light is *phos*. Atomists such as Democritus and Epicurus believed that all sensation was caused by direct contact of the objects with the organs of sense. Images of visible objects were conceived as material replicas of the objects streaming into the eye—the intromission theory of vision. On the other hand, Euclid, Hero of Alexandria, and others saw vision occurring by means of something discrete emanating from the eye and touching the seen objects, somewhat akin to a walking stick forming a tactile extension of the arm—the extramission theory.[1]

According to Aristotle's view—which endured for centuries—no material affluence existed either from the objects to the eye or in reverse, but rather vision ensued by dint of action on a material medium filling all space: "In fact, if the intervening space were void, not only would accurate vision be impossible, but nothing would be seen at all."[2] Light imparted on the substance of the medium a certain configuration, "of this substance light is the activity." Being a state of the medium rather than a substance within it, light required no time for its propagation, similar to a large expanse of water that may all simultaneously freeze. According to Ptolemy, light's velocity was instantaneous also because the moment we opened our eyes, we already saw the most distant stars.

The two Greek concepts of light as either a substance or an action upon a substance were revived during the Renaissance; René Descartes, L. Euler, T. Young, J. C. Maxwell, and others held to Aristotle's view of a material medium upon which light acted, the ether or aether:

Whatever difficulties we may have in forming a consistent idea of the constitution of the aether, there can be no doubt that the interplanetary and interstellar spaces are not empty, but are occupied by *a material substance or body* [emphasis added], which is certainly the largest and probably the most uniform *body* of which we have any knowledge.[3]

On the other hand, Newton and his school held that light itself was substance qualified by color: "For, since Colours are qualities of Light, having its rays for their intire and immediate subject, how can we think those Rays qualities also, unless one quality may be the subject of and sustain another; which in effect is to call it substance."[4] The predicament attended to a dual nature of light persisted until the end of the last century when—through the discoveries of Roentgen, Becquerel, the Curies, and others—it gradually emerged that the distinction between substance (mass) and action (energy) was not always clear, and subsequently, light was conveniently conceived—in the form of the wave mechanics theory—as both substance and action.[5]

The older conception of light as action that required no time to transfer from one position to another was based not simply on the authority of Aristotle, Ptolemy, Descartes, or Kepler but upon all available sensory (experimental) data procured by the technology of the time. In consequence, it was possible to think optics and solve optical problems with the help of Euclid's static geometry that gave this scientific endeavor an advantage over others by the certainty of its axioms and deductions and, hence, predictability. Euclid himself, in his *Optics*, took a strictly mathematical approach (except perhaps to his physical conception of visual rays); Aristotle, on the other hand, maintained that "optics investigates mathematical lines, but as being physical, not as being mathematical." The sensory and perceptual foundations of mathematics not being consciously recognized, light was comprehensible geometrically in terms of points in different positions and lines with different directions, but these lines did not have a certain width or length (representing a vector magnitude) as may nowadays be done.

A most fascinating and puzzling phenomenon is how does light from its source reach a distant substance, such as the eye, upon which it manifests itself? It is the mysterious problem of action at a distance, which applies also to gravity and magnetism. When a star explodes, does its gravity vanish instantaneously? And if not, what is the speed of gravity? Pierre-Simon Laplace (1749-1827) believed that universal gravitation was transmitted a million times more rapidly than light. Does gravity exert its effect in straight lines? When an electric current flows through a metallic wire, it generates magnetic effects, say, by changing the position of metal chips in the vicinity. When the current is increased, such as in a conventional Edisonian lightbulb, it may heat the wire to such a degree that it begins to emit light. Neither gravity or magnetism, nor light, can be known—are manifest and real—until they interact with some material substance though what a material substance is, is quite another question.

The problem of action-at-a-distance arose acutely from Newton's theories describing how gravity from the distant sun affected the rotating planets, or that from the earth affected the falling apple. "What *is* controversial is the idea of *unmediated* action at a distance, where there is both a gap between cause and effect and no intermediate causes and effects to fill it."[6] Some attempted to explain the phenomenon based on the theory of relativity and field theory: "By the term 'action at a distance' we mean relativistically invariant particle interaction."[7] But since relativity contains its own concepts of the nature of light, it is of no help.

The light from the sun, for instance, reaches us by traveling in empty space, in a void or vacuum, that must hence be full of unseen light. You don't see it there, just as you will not see the light of a candle in the center of the room unless you look directly at it or its reflection off the wall or off other substantial matter in the room, such as smoke or dust (Tyndall effect). Light from the sun traveling in empty space carries energy, but a few miles above earth, this space is extremely cold. When light encounters the substance of the earth, it heats it—days are warmer than nights, summers than winter. This action, generated by the momentum of light's impact, may contribute to the planets' circular motions that, in turn, by centrifugal forces, balance the centripetal force of gravity. The momentum of light was definitely demonstrated to affect small asteroids.[8] Does the light from the sun add to the weight of the earth and planets? As the sun's weight diminishes, does that of the planets' increase, elongating their orbits? The answer was easy in the days when all space was imagined filled with a material ether as Maxwell did:

> We have therefore reason to believe, from the phenomena of light and heat, that there is a aethereal medium filling space and permeating bodies, capable of being set in motion and transmitting that motion from one part to another, and of communicating that motion to gross matter so as to heat it and affect it in various ways . . . From these considerations Professor W. Thomson [Lord Kelvin] has argued, that the medium must have a density capable of comparison with that of gross matter, and has even assigned an inferior limit to that density.[9]

Newton began experimenting with light in 1666 and first published his "New Theory of Light and Colours" in 1672; whereas, Roemer did not discover that light was spatially and temporally successive until 1676. Newton thus conceived his *Opticks* in terms of static geometry; whereas, for his *Principia*, published in 1687, he found himself obliged to invent the calculus in order to better understand motion. By 1704, after the fact that light did move became established, it was easy for him to see why: When one moment light is intercepted on a screen that is positioned between the source and a wall, light disappears from the wall. When in the next moment the screen is removed, light reappears on the wall, i.e., light is successive in time.[10]

One and the same light, like anything in continuous motion, can be intercepted only once. As Heraclitus said, it is impossible to step twice into the same river; to which his colleague added that it was impossible to do so even once. It all depends on the duration of the step in reference to the velocity of the stream, which leads us back to uncertainty problems: there is no certainty that the light on the screen one moment is the same as that in the previous moment, although the probability is enormous. It is possible to step into the same river if one moves downstream with the speed of the water, but then, one cannot collect or intercept any water or fish flowing with it by holding a pail or a screen facing upstream.

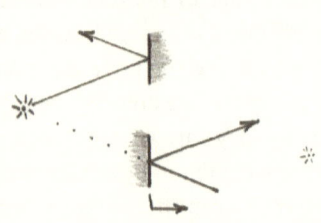

The velocity of light (or water) referred to a screen equals the velocity of a screen referred to that of light. The velocity of light referred to a screen may be increased when the velocity of light increases or when the screen is moving upstream, i.e., the approach or recession of an observer to the source equals the approach or recession of the source to the observer (see "Optokinematics"). The moment a stream of light is intercepted by a stationary screen, the light behind the screen, or the shadow, recedes with light's preexisting velocity. When the screen moves downstream with the velocity of light, no light is intercepted on either of its sides, and no shadow exists. When it moves in this direction faster than light, it overtakes it and intercepts light on the side facing downstream while its side facing the light source will be in shadow. In order to see a light source under these conditions, it is necessary to turn away from it, and the source appears in direction opposite to its real one (figure).

Under normal circumstances, the perception of light is invariably associated with the assumption that a source exists. No light has yet been seen without, sooner or later, confirming by other means the existence of its source. The smallest source of light has, since antiquity, been termed a *point source*, and light travels as beams (a semantic residue from the old stationary concept) where the thinnest beam is a ray, equivalent to the geometrical line. Similar as the geometrical concepts of the point and line evolved from real sensory data so do the optical point source and ray have real dimensions, albeit very small, depending on the chosen frame of reference.

The existence of light can be known only when received, and reception of light can occur only when the receptive entity—the eye, camera, or photoelectric element—is positioned within the direction of flow, thus to intercept it; light that passes by a screen can be perceived frontally but not laterally, from a position aside or behind the screen in the shadow. It distinguishes light from, say, electric flow in a conductor, which may be known laterally by means of the magnetic effects it engenders. This brings us to the task of measuring light, i.e., determining its quantity, a field named photometry.

For our purposes, in this volume, we generally presuppose an unchanging source, a known point source emitting light continuously, constantly, and uniformly—like a star in the sky. When speaking of magnitude, we shall, therefore, concern ourselves primarily with light already in existence, not with the mode of its generation. The magnitude of light, being an entity in motion, is given in the first place by the magnitude of this motion—the velocity. Motion in a straight line, like the length of the line we discussed earlier, can be determined only from outside the line, i.e., by a lateral view. Since light cannot be received laterally, and one and the same light cannot be received frontally, it follows that theoretically one cannot measure the velocity of one light—though practically, the point is mute as pointed out in the river example.

Light as a physical entity possesses a second dimension in addition to the one of the geometrical line or ray that represents its *direction* or *vector quantity*. The beam possesses some two-dimensional magnitude of *cross section* perpendicular to the direction of flow. When light is represented in our illustrations as a line, it denotes principally its direction only. For a given distance from a given point source, the total quantity of light is proportional to the size of the area; a larger area receives more light. Quantity per area is termed concentration or density. When the same quantity of light present on a large area is dispersed over a smaller one, the density per unit area increases, and vice versa.

The nature of this proportionality does not seem subject to simple mathematical deduction, i.e., double the area does not necessarily lead to half the prior quantity of light per unit area, and hence, the density may have to be determined empirically.[11] The predicament arises from the absence of units for light in cross section, for light is not a material substance. There exists no minimum unit (molecular, atomic, or nuclear) for light in cross section. It is possible to measure the velocity of something without going into the nature of the thing because velocity is in reference to material objects. But in order to measure the thing itself in concrete units of length and area, we need know what it is, a problem encountered also in connection with heat and gravity.

A ray of light, like the geometrical line, has only length and no width; it therefore has no sides. When new optical phenomena were discovered in the seventeenth century, such as double refraction, they could not be understood in terms of unidimensional rays, and the need arose for the invention of a three-dimensional material medium—the aether—to accommodate them. Ever since, the science of optics appeared quasi-schizophrenic—it seemingly possessed two lines of logic, one geometrical (metaphysical), the other physical—geometrical optics and physical optics.

When light's motion is interrupted, say, by a screen, it causes an effect. The quantity of this effect is termed brightness, luminance, illuminance, or illumination hereafter termed intensity in this narrow sense (in reference to the power of the emitting source the term *luminous intensity* will be specifically denoted). Brightness or intensity is not light itself but an effect produced by light when interrupted—an effect manifested in the eye, on a photographic film, or by a photoelectric cell. When hit by a falling stone, that's not gravity but its effect.

The distinction between cause and effect is difficult because we know light only through its effects; and only gradually, by shedding off these effects, does a clearer picture of its true nature emerge. For example, as far as we know, matter is discontinuous; it is formed of discrete particles such as atoms. When light interacts with matter (say, by reflection, refraction, or diffraction) the effect on light is often discontinuous, *periodic*; but this does not necessarily mean that light itself is periodic.

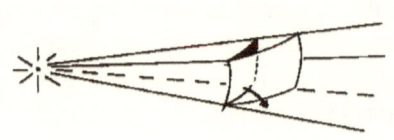

Since it appears impossible to directly determine the density of light in a beam, we resort to observing the effects it produces. The effect of light, its intensity, depends on area size. According to Kepler (reasoning corpuscularly), a point source emits light in all directions so that the source forms the center of a sphere (figure). The surface area of a sphere relates to the square of the radius, and therefore, from a point source, the intensity of light that is received per surface area perpendicular to the direction of flow diminishes as the square of the area's distance from the source. For a beam of rays traveling parallel to each other, coaxially, the cross-section area is minimum when cut perpendicular to the direction of flow; and hence, the intensity at perpendicular incidence is maximum. It declines with *obliquity* to this direction.[12]

Light being in motion, its quantity that crosses a given area per unit time depends in the first place on light's velocity; the slower the light, the less of it crosses a given stationary area per unit time. The intensity diminishes when the velocity of light diminishes—as when it travels through transparent media—but it also diminishes when the area is increased—as when the cross section is not perpendicular to the direction of flow. This fact is quintessential to optokinetic understanding of light: the effect (the brightness or intensity) covaries both with the velocity or with area size. The fact of changed intensity does not provide decisive information about its cause, which may lie in either a changed density—variability in space—or changed velocity—variability in time. The decision as to which causes the change in the effect hinges on empirical measurements of one or the other. This duality also means that when a phenomenon is a function of the intensity, it will vary in the same direction whether area size varies or the relative velocity.

Leafing through the annals of photometry, one begins to appreciate the difficulties on the road to forming agreed concepts and units concerning the quantity of light. As Perry Moon and D. E. Spencer put it, "The concepts, names, and units of photometry have never been thoroughly standardized; considerable variation in nomenclature is to be found in the literature."[13] It was easier to agree on a standard emitting source—such as a candle or a given star—than to standardize the effect, the intensity. Light was conceived as a wave motion of the medium—or a wave motion of particular quantities (quanta)—where no correlation existed between the length or frequency of the waves and intensity. An unknown source of light may be compared to a standard source by having the lights fall side by side on a screen. When the lights effected different color sensations, and were therefore supposed to consist of different waves, no firm correlation existed; for it is almost impossible to compare intensities of, say, red to those of white or green without a common denominator.

The effect of yellow light on a green plant differs from its effect on a blue fish or its effect on mercury varies from that on selenium. It is, of course, possible to set the effect on one particular entity as a standard, say selenium, and measuring other effects in comparison to this one as was done with heat and mercury. Heat differs, however, from light; for it was understood in a single term—motion—whereas light was conceived as innumerable wavelengths. The effect of a little red may equal that of a larger quantity of violet. Units of magnitude determined by comparison to selenium will be of little utility when attempting, for instance, to decide on the level of light in a room.

Since Newton, the effect of inducing color sensations was believed already present in its cause, in the light itself—"And colors are affirm'd to be not qualifications of Light, deriv'd from Refractions of natural Bodies, [as 'tis generally believed;] but Original and Connate properties, which in divers rays are divers."[14] In addition, white light was seen as an aggregate of colored lights, and all these lights traveled in free space with equal velocities: "At all events we know with great exactness that this velocity is the same for all colours, because if this were not the case, the minimum of emission would not be observed simultaneously for different colours during the eclipse of a fixed star by its dark neighbour."[15] With these ideas deeply ingrained in the mind, a meaningful measure of intensity remained elusive.

The one feature of light that does not depend on its effect—the velocity—is a terrific magnitude, almost beyond conception because one never perceives anything like it. When a light is lit, it travels in air or vacuum a distance equal to rounding the earth ten times (in New York latitude) by the time one finished saying "one." The requisite technology for the inclusion of light's velocity into consideration of ordinary optical and photometric problems is perhaps inadequate to the task. Furthermore, there is little incentive to attempt develop this technology as long as one embraces the belief that this velocity is forever constant.

> Very lame and imperfect theories are sufficient to suggest useful experiments, which serve to correct those theories, and give birth to others more perfect. These, then, occasion farther experiments, which bring us still nearer to the truth; and in this method of *approximation*, we must be content to proceed, and we ought to think ourselves happy, if, in this slow method, we make any real progress.[16]

REFERENCES

1. Lindberg, D. C.: Theories of Vision. Univ. of Chicago, 1976.
2. Aristotle: DeAnima 2.7; 419, 12-22, in: The Works of Aristotle Translated into English (Smith, J. A., Ross, W. D., ed.), Oxford, 1908-1952.
3. Maxwell J.C. "Ether", in *Encyclopaedia Britannica, Vol. 8; 1878: 568-72.*
4. Newton, I.: New Theory about Light and Colors. *Phil, transact.* 80; 3085, 1672.
5. Mark H.H. Physick and Physics. *Conn. Med. 69(4), 2005: 247-8.*
6. Poidevin R.L. Action at a distance, In: Philosophy of Science, Vol. 61 (A. O'Heear, ed.) Cambridge Univ. Press, 2007: 21-36.
7. Hoyle E., Narlikar, J.V.: Action at a Distance in Physics and Cosmology. WH Freeman, New York; 1974:97-98.\
8. Lowry C.L. et al: Direct Detection of Asteroidal YORP effect. *Science* 316 (5822); 2007: 72-4.
9. Maxwell J.C.: A dynamical theory of the electromagnetic field. *Transact. Roy. Soc.* London; 1865:459-512.
10. Newton, I.: Opticks. New York, Dover, 1952, p. 2.
11. Mach, E.: The Principles of Physical Optics. New York, Dover, 1962, pp. 17-20.
12. Born, M., Wolf, E.: Principles of Optics. 3rd ed., New York, Pergamon, 1965, p. 182.
13. Moon, P., Spencer, D. E.: Photometry, in: Encyclopedia Britannica, Chicago, Wm. Benton, 1967, vol.17, p. 991.
14. Newton, I.: op. cit. ref. 4.
15. Einstein, A.: Relativity. New York, Crown, 1961, p. 17.
16. Priestley J. The history and present state of discoveries relating to vision, light, and colours. London, J Johnson; 1772: 181.

Index